37.75
63R

W9-BBU-379

CRITERION-REFERENCED TEST DEVELOPMENT

HF
5549.5
.T7
5554
1989

GP9 0 00992

CRITERION-REFERENCED TEST DEVELOPMENT

Technical and Legal Guidelines for Corporate Training

SHARON A. SHROCK

WILLIAM C. C. COSCARELLI

ADDISON-WESLEY PUBLISHING COMPANY INC.

Reading, Massachusetts   Menlo Park, California
New York   Don Mills, Ontario   Wokingham, England
Amsterdam   Bonn   Sydney   Singapore   Tokyo
Madrid   San Juan

Library of Congress Cataloging-in-Publication Data

Shrock, Sharon A.
  Criterion-referenced test development : technical and legal guidelines for corporate training / Sharon A. Shrock, William C. C. Coscarelli

    p.   cm.
  Bibliography: p.
  Includes index.
  ISBN 0–201–10328–1
    1. Employees—Training of—Evaluation.   2. Criterion-referenced tests.
I. Coscarelli, William C. C.   II. Title.
HF5549.5.T7S554   1989
658.3′12404—dc20

Copyright © 1989 by Addison-Wesley Publishing Company, Inc.

All rights reserved. No part of this publication may be reproduced, stored in a retrieval system, or transmitted, in any form or by any means, electronic, mechanical, photocopying, recording, or otherwise, without the prior written permission of the publisher. Printed in the United States of America. Published simultaneously in Canada.

Cover design by Hannus Design Associates
Text design by Anna George
Set in 10-point Bookman Light by Modern Graphics, Inc., Weymouth, MA

ABCDEFGHIJ-MA-89
First printing, August 1989

To Rubye and Don
and to Kate

Table of Contents

Preface

Accounts of the explosion of activity and investment in human resources development and corporate training are everywhere. Fueled by rapidly changing technology, worldwide economic forces, and the crisis in American education, the increases in expenditures show no signs of abating. Partly as a result of so much investment, a greater emphasis on accountability would seem predictable. Instructional designers will increasingly be called upon to document the results of their work with "hard data." Enter into this already complex scenario an increasingly litigation-oriented consumerate and workforce. At a time when more testing and more careful certification of personnel would seem warranted, the pressures for *excellence* in testing have never been greater.

The nation's HRD and instructional design professionals face an enormous challenge in the testing area. Their expertise in the systematic creation of instruction has grown tremendously in the last decade, yet knowledge of the technology of testing has lagged behind. It is our hope that this book will help these professionals meet the testing demands that lie ahead of them.

The content of this book was drawn largely from a testing workshop that we have delivered for instructional designers representing a broad range of companies, agencies, and educational institutions large and small. In this regard, the content has already been field tested for relevance and suitability in an applications-oriented environment. We have tried to balance the rigor of psychometrics with the demands placed upon practicing instructional designers.

It is also probably worth noting that this book was written by two instructional designers, and we would like to think that that has shaped the presentation of its content. Decisions about sequencing the content were the most difficult. The test design process is described in a more linear fashion than its implementation is likely to be. Most instructional designers are, however, by now accustomed to seeing static-looking models that represent a whirlwind of iterative activity, so we expect to be forgiven by many. The redundancies in the book are intentional attempts to make the book useful as a reference for those who prefer to use it that way. Every effort has been made to make statistics calculations doable for anyone with a hand-held calculator, and more importantly, understandable for everyone.

Many have played a role, knowingly or otherwise, in the creation of this book. We would like to thank Dr. Kenneth Stanley Majer (in whose psychometrics class we met one another) for his commitment to a conceptual understanding of testing. We would also like to acknowledge Ramesh, Sam, Bob, and Lorie for the opportunity to develop the initial corporate Criterion-Referenced Test Development workshop, as well as their colleagues for the refinement of the content that forms the foundation for the book. Finally, a word of thanks to the National Society for Performance and Instruction for having hosted the CRTD workshop for several years.

Sharon A. Shrock
William C. C. Coscarelli
Carbondale, Illinois

Introduction: A Little Knowledge Is Dangerous

WHY READ THIS BOOK?
A CONFUSING STATE OF AFFAIRS
WHAT IS TO COME . . .

WHY READ THIS BOOK?

Corporate training, driven by competition and a sense of "the bottom line," has a certain intensity about it. Errors in instructional design or failure to master content can result in significant negative consequences. It's not surprising, then, that corporate trainers are strong proponents of the systematic design of criterion-referenced instructional systems. What is surprising is the general lack of support for a parallel process of assessment of instructional outcomes—in other words, testing.

Most training professionals have taken at least one intensive course in the design of instruction, but most have never had similar training in the development of criterion-referenced tests—tests that compare people against a standard, instead of against other people (norm-referenced tests). It's not uncommon for a 40-hour workshop in the systematic design of instruction to devote less than four hours to the topic of test development—focusing primarily on item writing skills. With such minimal training, how can we make and defend our assessment decisions? For example, how can you show that those graduates you certify as "masters" are indeed masters and can be trusted to perform competently while installing, say, an expensive switching device? And what would you tell an EEO officer who pre-

1

sents you with a grievance from an employee who was denied a salary increase based on a test you developed?

Without an understanding of the basic principles of test design you can face difficult ethical, economic, or legal problems. For these and other reasons, test development should be on an equal footing with instructional development—if not, how will you know whether you got where you were going?

A CONFUSING STATE OF AFFAIRS

Grade schools, high schools, universities, and corporations share many reasons for not adopting the techniques for creating criterion-referenced tests. We have found three reasons that seem to explain why those who might otherwise embrace the technique haven't: misleading familiarity, inaccessible technology, and procedural confusion. In each instance, it seems that a little knowledge about testing has proven dangerous to the health of the criterion-referenced test.

Misleading Familiarity

As training professionals, few of us teach the way we were taught. However, most of us are still testing the way we were tested. We all took many tests while in school, and are familiar with them. Therefore, we tend to believe that we all know how to construct a good test. This belief is wrong, not only because exposure does not guarantee know-how, but because most of the tests we were exposed to in school were poorly constructed. The exceptions—the well-constructed tests in our past—tend to be the group-administered standardized tests, e.g., the Iowa Tests of Basic Skills or the Scholastic Aptitude Tests. Unfortunately for corporate trainers, these standardized tests are good examples of norm-referenced tests, not criterion-referenced tests. Norm-referenced tests are designed for completely different purposes than are criterion-referenced tests, and both are constructed and interpreted differently. Most teacher-made tests are "mongrels," having characteristics of both norm-referenced and criterion-referenced tests—to the detriment of both.

Inaccessible Technology

Criterion-referenced test technology is scarce in corporate training partly because the technology of creating these tests has been slow to develop. Even now with so much emphasis on minimal competency testing in the schools, most college courses on tests and measurements are about the principles of creating norm-referenced tests. In other words, even if trainers want to "do the right thing," answers to important questions are hard to come by. Much of the information about criterion-referenced tests has appeared only in highly technical measurement journals. The technology to improve practice in this area just hasn't been accessible.

Procedural Confusion

A third pitfall that we have found in good criterion-referenced test development is that both norm-referenced tests and criterion-referenced tests share some fundamental measurement concepts, such as reliability and validity. Test creators don't always seem to know how to modify these concepts in order to apply them to the two different kinds of tests.

Recently we saw an article in a respected corporate training publication that purported to detail all the steps necessary to establish the reliability of a test. The procedures that were described, however, will work only for norm-referenced tests. Since the article appeared in a training journal, we question the applicability of the information to the vast majority of testing that its readers will conduct. As the author was the head of a training department, we had to appreciate his sensitivity to value of a reliability estimate in the test development process, yet the article provided a clear illustration of procedural confusion in test development, even among those with some knowledge of basic testing concepts.

WHAT IS TO COME . . .

In the following chapters we describe a systematic approach to the development of criterion-referenced tests. Part I, The Fundamentals, provides a basic frame of reference for the entire test development process. Part II, Planning and Creating the Test, describes the use of a course hierarchy in planning a test and presents guidelines for the technical aspects of item writing, the determination of test length, and the construction of different kinds of rating scales. Part III, Piloting the Test, is a discussion of the formatting, organization and administration of a test. It includes procedures for collecting initial formative evaluation test data to be used for test improvement. Part IV, Evaluating the Test, describes item analysis techniques useful for choosing and improving test items and procedures for determining the cut-off score that defines mastery. How to establish the reliability and validity of both performance and paper-and-pencil tests is also included. Part V, Legal Issues in Criterion-Referenced Testing, explores some of the important legal issues that surround this topic and presents procedures for determining the possible "adverse impact" of tests.

Periodically, we have provided an opportunity for practice and feedback. You will find that each of the topics in Part I is reinforced by exercises with corresponding answers (because we feel the fundamentals are fundamental), and that throughout the book opportunities to practice applying the most important or difficult concepts are similarly provided.

PART ONE

THE
FUNDAMENTALS

1.

Test Theory

WHAT IS TESTING?
WHAT DOES A TEST SCORE MEAN?
RELIABILITY AND VALIDITY: A PRIMER

WHAT IS TESTING?

There are four terms that are, at first, somewhat confusing: *testing, measurement, assessment,* and *evaluation.* These terms are sometimes used interchangeably; however, we think it is useful to make the following distinctions among them:

- *Testing* is the collection of quantitative (numerical) information about the degree to which a competence or ability is present in the test taker. There are right and wrong answers to the items on the test, whether it be a test comprised of written questions or a performance test requiring the demonstration of a skill. A typical test question might be, "List the six steps in the selling process."

- *Measurement* is the collection of quantitative data to determine the degree of whatever is being measured. There may or may not be right and wrong answers. A measurement inventory such as the *Decision Making Inventory* might be used to determine a preference for using a systematic style versus a spontaneous one in making a sale. One style is not "right" and the other "wrong;" the two styles are simply different.

7

- *Assessment* is systematic information collection without reference to making judgments of worth. It may involve the collection of qualitative (narrative) as well as quantitative information. For example, through the use of a series of personality inventories and interviewing, one might build a profile of "the aggressive salesperson." (Many companies use assessment centers as part of their management training and selection process. However, as the results from these centers are usually used to make judgments of worth, they are more properly classed as evaluation devices than as assessment devices.)

- *Evaluation* is the process of making judgments regarding the appropriateness of some person, program, process or product for a specific purpose. Evaluation may or may not involve testing, measurement, or assessment. Most informed judgments of worth, however, would likely require one or more of these data-gathering processes. Evaluation decisions may be based on either quantitative or qualitative data; the type of data that is most useful depends entirely on the nature of the evaluation question. An example of an evaluation issue might be, "Does our training department serve the needs of the company?"

PRACTICE

Here are some statements and questions related to these four concepts. See if you can classify them as issues related to testing, measurement, assessment, or evaluation.

1. "She was able to install the air conditioner without error during the allotted time."

   _____

2. "Personality inventories indicate that our programmers tend to have higher extroversion scores than introversion scores."

   _____

3. "Does the pilot test process we use really tell us anything about how well our instruction works?"

   _____

4. "What characteristics should we use to select submarine officers?"

FEEDBACK

1. Testing
2. Measurement
3. Evaluation
4. Assessment

WHAT DOES A TEST SCORE MEAN?

Suppose you had to take an important test. In fact, this test was so important that you studied intensively for it for five weeks. Suppose then that when you went to take the test, the temperature in the room was 45 degrees. After 20 minutes all you could think of was getting out of the room, never mind taking the test. Suppose, on the other hand, you never studied for the test. By chance a friend dropped by the morning of the test and showed you the answer key. In both situations, the score you received on the test probably didn't accurately reflect what you actually knew. In the first instance, you may have known more than the test score showed, but the environment was so uncomfortable that you couldn't attend to the test. In the second instance, you probably knew less than the test score showed, due now to another type of "environmental" influence.

In either instance the score you received on the test (your observed score) was a combination of what you really know (your true score) and those things that might modify your true score (error). This relationship is the basis for all test theory. It is usually expressed by a simple equation:

$$Xo = Xt + Xe$$

where Xo is the observed score, Xt the true score, and Xe the error component.

It is very important to remember that, in test theory, "error" doesn't mean a wrong answer. It means any mismatch between a test-taker's actual level of knowledge (the true score) and the test score the person receives. Error can make a score higher (as we saw when our friend dropped by) or lower (when it got too cold to concentrate).

The primary purpose of a systematic approach to test design is to reduce the error component so that the observed score and the true score are as nearly identical as possible. All the procedures we will discuss and recommend are tied to a simple assumption: The primary purpose of test development is the reduction of error. In terms of our equation, we think of the results of test development like this:

$$Xo = Xt + xe$$

where error has been reduced to the lowest possible level.

Realistically, there will always be some error in a test score, but careful attention to the principles of test development and administration will help reduce the error component.

PRACTICE

See if you can list at least three situations that could inflate a test-taker's score and three that could reduce the score.

Inflation Factors	*Reduction Factors*
1. Sees answer key	1. Room too cold
2. _____	2. _____
3. _____	3. _____
4. _____	4. _____

FEEDBACK

Inflation Factors	*Reduction Factors*
1. Sees answer key	1. Room too cold
2. Looks at someone's answers	2. Test scheduled too early
3. Unauthorized job aid used	3. Noisy heating system in room
4. Answers are cued in test	4. Can't read test directions

RELIABILITY AND VALIDITY: A PRIMER

Reliability and validity are the two most important character-istics of a test. Later on we are going to explore these topics and provide you with specific statistical techniques for determining these qualities in your tests. For now, we simply want to provide an over-view so that you will see how these ideas serve as standards for our attempts to reduce error in testing.

Reliability

Reliability is the consistency of test scores. There is no such thing as validity without reliability, so we want to begin with this idea. There are two kinds of reliability that are typically considered in criterion-referenced test construction: test-retest and inter-rater.

- *Test-retest reliability* is the measure of an individual's test score consistency over time. In other words, did the test-takers get the same scores on a second administration of the test as they did on the first (assuming no practice or instruction occurred be-tween the two administrations and the administrations were relatively close together)? If each test-taker received the exact same score the second time as the first, then you have perfect reliability. If there is no relationship between the test scores, then you will have a reliability estimate of zero.

- *Inter-rater reliability* is the measure of consistency among in-dividual judges' ratings of a performance. If you have determined that a performance test is required, then you need to be sure that your judges (raters) are consistent in their assessments. In Olympic competition we expect that the judges' scores should not significantly deviate from each other. The degree to which they agree is the measure of inter-rater reliability. This agree-ment will also vary between zero and perfect.

Validity

Validity has to do with whether or not a test measures what it is supposed to measure. A test can be consistent (reliable) but mea-sure the wrong thing. For example, assume that we have designed a course to teach employees how to install a new telephone switchboard. We could devise an end-of-course test that asks learners to list all the steps for installing the new equipment. We might find that the learners can consistently list these steps, but that they can't install the switchboard, which was the intended goal of the course.

FIGURE 1.1.A Reliable, but not Valid

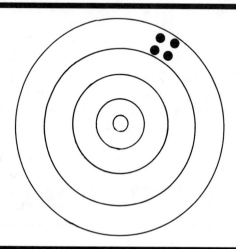

FIGURE 1.1.B Reliable and Valid

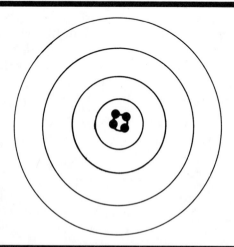

FIGURE 1.1.C Neither Reliable nor Valid

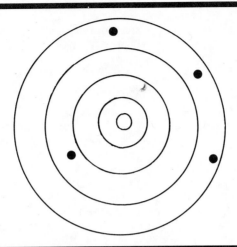

Hence, our test is reliable, but not a valid measure for the installation task.

Figure 1.1 illustrates the relationship between reliability and validity. In Figure 1.1a, the marksman has fired all of her shots in a tight group. Her shooting might be termed "reliable" because the shots are all in the same place, but her shooting isn't valid since she missed the bullseye. In Figure 1.1b, her shots are reliable and valid; she consistently hit the bullseye. In Figure 1.1c, the shots are neither reliable nor valid. Notice that it is not possible for the marksman's shots to be valid without also being reliable. Hence, the truism that a test cannot be valid if it is not reliable.

PRACTICE

1. "Bob, I don't know if these tests should be considered as reliable measures of performance. What do you think?"

Person	Week 1 Score	Week 2 Score
Sid	89	90
Diane	92	90
Michelle	75	79
George	65	68

2. "Lorie, here's the test you wanted to see. We selected the items to match the job descriptions for our participants. The test scores are highly reliable from one test administration to the next. Do you think this will work?"

FEEDBACK

1. The test appears to be reliable. The scores are very close between each administration. The time lapse of one week is probably a good choice. Waiting too long encourages forgetting or additional learning of the content; not waiting long enough allows pure memorization of the test items.

2. The test may well be valid. The items are linked to the job descriptions, which should increase the likelihood that the items are valid measures of expected performance. Furthermore, the test has demonstrated reliability, a prerequisite for validity. However, it would be impossible to know for sure whether the test was valid without conducting a validation study as described in Chapter 11.

2.

Types of Tests

CRITERION-REFERENCED VERSUS NORM-REFERENCED TESTS

There are two major ways to interpret test scores: criterion-referenced interpretation and norm-referenced interpretation. While some tests can be interpreted both ways, this is usually not the case. Tests should be constructed in order to facilitate either a criterion-referenced or a norm-referenced interpretation. Basically, norm-referenced tests need to be composed of items that will separate the scores of test-takers from one another, while criterion-referenced tests need to be composed of items based on specific objectives, or competency statements. To understand this difference in test interpretation it is helpful to begin by understanding frequency distributions—graphic representations of test scores.

Frequency Distributions

Figure 2.1 shows a frequency distribution for 25 people who have taken a test. The range of possible test scores is listed on the horizontal axis; the number of people who might obtain a given score is listed on the vertical axis. In this figure, five people scored 50,

15

FIGURE 2.1 Frequency Distribution

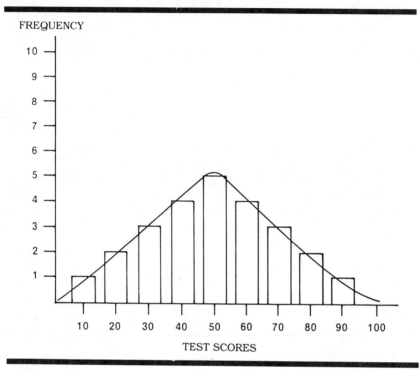

two people scored 20 and 80, while nobody scored 100. These points are connected to create a smooth curve.

Norm-Referenced Test Interpretation

A norm-referenced test (NRT) interpretation defines the performance of test-takers in relation to one another. If you want to rank people for the purpose of selecting the top performers, your ideal frequency distribution would look like Figure 2.2, because, in this situation, each test score was attained by only one person. The ranking of these test-takers would be easy because there are no tied ranks, i.e., no two test-takers got the same score.

Now, you will rarely have a test that separates everyone quite so perfectly. Instead, most NRTs will have distributions that look like Figure 2.3. Figure 2.3 is the classic shape of the NRT distribution;

FIGURE 2.2 Ideal Frequency Distribution for Ranking

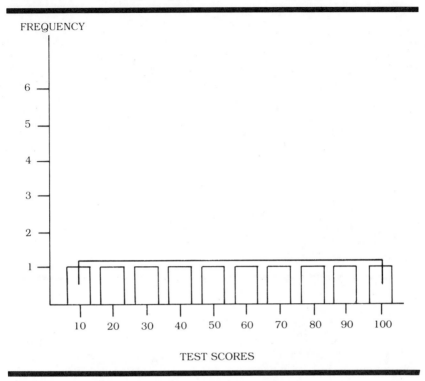

you may have heard it called the "bell curve" or the "normal distribution."

A normal distribution is what typically results from the administration of an NRT. Unlike our ideal in Figure 2.2, people will tend to cluster in the middle ranges. However, the scores still represent a wide spread, i.e., the test scores have been successfully separated from one another. This spread of scores is good for NRT interpretation because it increases the confidence with which we can decide that one test-taker scored better than another. And the comparison of test-takers to one another is what NRT interpretation is all about.

Norm-referenced tests can be very useful. Medical schools use the Medical College Aptitude Test (MCAT) to help predict success in medical school. Because of the large number of people applying for medical school and the limited number of openings, medical schools have chosen to use the MCAT as one way of ensuring that the best students are admitted. (As a patient, you would probably prefer to

FIGURE 2.3 Typical Norm-referenced Test Frequency Distribution

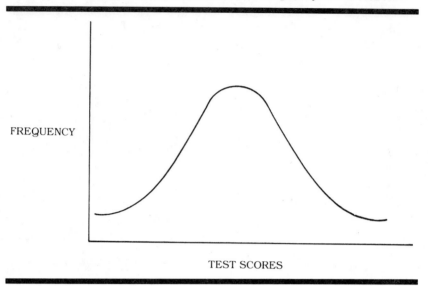

know that only the best students are being admitted.) Norm-referenced tests are ideal for making this kind of selection decision, when we must choose the best test-takers among a group.

Criterion-Referenced Test Interpretation

In contrast to the NRT, the criterion-referenced test (CRT) defines the performance of each test-taker without regard to the performance of others. Unlike the NRT where success is defined in terms of being ahead of someone else, the CRT interpretation defines success as being able to perform a specific task or set of competencies. There is no limit to the number of people who can succeed on a criterion-referenced test, unlike the NRT. Very often the CRT frequency distribution looks like the one in Figure 2.4. This distribution is often called a "mastery curve."

Unlike NRTs, the shape of the CRT distribution is not essential to its interpretation, since the scores are interpreted in terms of the competencies the scores represent rather than in terms of the scores' relationship to one another. The reason that a CRT frequency dis-

FIGURE 2.4 Typical Criterion-referenced Test Frequency Distribution

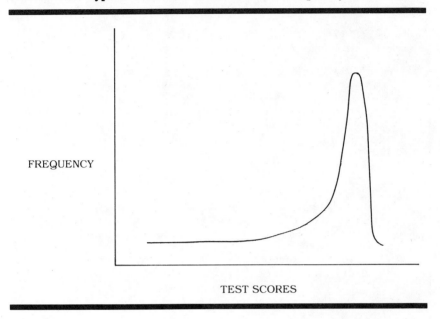

tribution often looks like Figure 2.4 is that the test items are based on specific competencies, and the instruction that the test-takers receive in anticipation of the test is usually addressed specifically to these competencies. Therefore, many test-takers do well on the CRT, resulting in a distribution in which most test-takers are clustered near the high end.

Criterion-referenced tests should be used whenever you are concerned with assessing a person's ability to demonstrate a specific skill. The medical boards' licensing exams are examples of tests whose philosophy is criterion-referenced. If you are being operated on, you should know that your surgeon is competent to perform the operation, not just that he or she is better than 90% of those who graduated. Merely knowing more than the others in the class doesn't guarantee that your surgeon can perform the operation; maybe nobody in the class mastered the operation. The danger of using NRTs in corporate training situations is that, because the tests do not refer to specific competencies, what test-takers can actually do is unverifiable.

PRACTICE

Here are two scenarios; decide which type of test—norm-referenced or criterion-referenced—is required for each one.

1. Many foreign countries administer tests as early as eighth grade to determine who among test-takers will be eligible to attend schools of different types, e.g., vocational, engineering, medical, etc. Scores on these tests correlate with subsequent achievement in advanced classes. There is never enough space in each of the schools for all of the students who want to attend.

_____

2. A number of years ago England developed a national program called the Open University. By using television and textbooks keyed to objectives, anyone in the country can take instruction at home that will lead to a degree from the University. Students must master a given percentage of the objectives for the courses they take.

_____

FEEDBACK

1. Norm-referenced testing
2. Criterion-referenced testing

FIVE PURPOSES FOR TESTS IN INSTRUCTIONAL SETTINGS

Before we construct a CRT, we need to know what the purpose of our test will be in the instructional setting. There are five basic purposes for criterion-referenced tests:

- *Prerequisite tests* are used to ensure that the learners have the background knowledge required for success in the course. If there are minimum skills required for the course, the prerequisite test is designed to assess mastery of these skills.

- *Entry tests* are used to identify the skills to be taught in the course that the entering student may already possess. The entry test can be used to allow students to bypass a module of instruction, if the students demonstrate in advance the skills to be covered in the module. This test can also be used to identify the range of skills the students have, over and above the prerequisite skills.

- *Diagnostic tests* are used to assess mastery of a group of related objectives in an instructional unit. Whereas entry and prerequisite tests are used before instruction, the diagnostic test will typically be used during instruction (when it is often called an "embedded test") or as part of the post-test process to determine exactly where a learner is having difficulty.

- *Post-tests* are administered after instruction to assess the test-taker's mastery of terminal objectives, i.e., end-of-course objectives.

- *Equivalency tests* are used to determine whether a learner has already mastered the course's terminal objectives without going through instruction. These tests are used to determine if a test-taker can bypass—"test out of"—an entire course.

You will find that often the same questions will appear on different types of tests. For example, the items on the post-test may be the same as, or similar to, the items on the equivalency test. In fact, all of these tests are relative—a prerequisite test for one course could be a post-test for a previous course. The type of test is determined by the purpose the test serves—not by the text of the items it contains.

PRACTICE

Here are five quotes. Which type of CRT would you recommend for each situation?

1. "Look, I already know this stuff. There is no need for me to travel to headquarters, lose time in the field that I can use for sales, and take some irrelevant class."

   _____

2. "I've got to teach a class of 30 people from all over. I have no idea who knows what."

   _____

3. "Well, you've been through five days of training; now let's see how you do."

   _____

4. "I think I have the general idea about how to install the switch-board, but there seem to be a couple of areas I can't quite do right."

_____

5. "If I'm going to have 30 people in each of my classes, then we really need to make sure they have all the basic skills they will need to get through without always asking me for help."

_____

FEEDBACK

1. Equivalency

2. Entry

3. Post-test

4. Diagnostic

5. Prerequisite

THREE METHODS OF TEST CONSTRUCTION (ONE OF WHICH YOU SHOULD NEVER USE)

There are three basic methods for constructing a test: topic based, statistically based, and objectives based. You should never use the topic based approach.

Topic Based Test Construction

The topic based approach is the way in which most tests are created. Almost all teacher education classes recommend this strategy, and it is probably the only approach commonly used in most classrooms today. Essentially, the instructor takes a given number of questions from, say, Chapter 3 or Topic 1, or some number of questions from Chapter 4 or Topic 2, loosely basing the number of questions on the perceived importance of the topic. This practice is imprecise and certainly doesn't allow for criterion-referenced interpretation; usually the test distribution is not widespread enough to allow for norm-referenced interpretation either. In other words, topic based construction frequently will not separate test-takers, as is required for the NRT, nor will it verify specific competencies as does the CRT.

Statistically Based Test Construction

Norm-referenced tests are constructed through use of a statistically based method. This means that items are chosen that have been shown to separate test-takers. The Scholastic Aptitude Tests separate test-takers and can be used to predict some types of success in college. However, the SATs don't define specifically what is measured in the precise way required for criterion-referenced interpretation.

If you were constructing an NRT, you would look for items that separate people for whatever reason along the dimension of interest to you. Consider this: In one nationally recognized NRT the test-taker is asked "Who do you think was a better president—Washington or Lincoln?" If a test-taker answers "Washington" the response will support his or her classification as "abnormal." This basis of classification has nothing to do with an objective criterion of being abnormal, i.e., the item does not specify clinically abnormal behaviors against which the test-taker's behavior can be judged. Rather, the classification is warranted because people who have been classified as "abnormal" due to verifiable behaviors tend to select "Washington" over "Lincoln" when given this item. If the relationship between abnormal behavior and the preference for "Washington" is strong enough, preferring "Washington" (along with "abnormal" responses to other such items) is enough to make the classification. Finding a group of such items that separate people is the heart of NRT development. Since this book is about the creation of criterion-referenced tests, we will not discuss these statistically based item selection methods in great detail.

Objectives Based Test Construction

In contrast to the NRT construction process, a CRT is based on items that assess a specific competency. Most corporate training philosophies specify an objectives based system. After all, most airline companies, and passengers, want to know that the pilot can land the plane, not just that he was better than everyone else in the class. Look at Figure 2.5. Notice how the test item follows directly from the behavior specified in the objective. With the "Washington/

FIGURE 2.5 Example of an Objective and Matching Test Item

OBJECTIVE: Given a selection of geometric figures, the student will be able to identify a previously unseen triangle.

      TEST: Which of the following is a triangle?

              ◯   △   ☐   ▱

Lincoln" example item above, the test item doesn't logically follow from any known description of clinically abnormal behavior.

To sum up the process of test construction and interpretation, just remember:

> "Topics are taboo; Statistics are normal, but Objectives meet the criterion."

In other words, avoid topic based test construction, use a statistically based approach to develop NRTs, and an objectives based approach to develop CRTs.

PRACTICE

Here are three scenarios. Read each and classify it according to the test construction process it describes.

1. "The new switching system is critical to our success with this project. We can't afford to have any errors with its installation. We need to make absolutely certain that our trainees can perform this task quickly and flawlessly."

   _____

2. "We have a lot of candidates for the position of shuttle pilot. Before investing in the expensive training of those few who will become pilots, we need to select carefully those with strong general analytical abilities."

   _____

3. "Let's see, we've covered a lot of territory in the workshop. I thought the stuff on management styles was well done, so I'll put about five questions on that, another three or four might go to performance appraisal techniques . . ."

   _____

FEEDBACK

1. Criterion-referenced test construction
2. Norm-referenced test construction
3. Topic based test construction

PART TWO

PLANNING AND CREATING THE TEST

3.

Hierarchical Analysis

HIERARCHIES

Learning theorists have found that many learning goals can be thought of as hierarchical in nature, i.e., subordinate skills are prerequisites to the final task. More specifically, some of the assumptions of this approach are

- a final goal can be analyzed into component skills that are quite distinct from each other; and
- the component skills are mediators of the final goal, i.e., mastery of a lower level skill is necessary to achieve the next level of performance; nonmastery of a subordinate skill significantly reduces the probability that the next level task will be mastered.

A hierarchical analysis can be approached from a variety of perspectives to include both mental and physical performances on the job. This chapter assumes that you are skilled in the techniques of job analysis in general, and offers a way of thinking about how to use a hierarchical analysis of tasks for the purpose of efficient testing.

An analysis of the course content to be tested, or more significantly, the content of a job where an employment decision may be made, is absolutely critical to the testing process. In the *Standards for Educational and Psychological Testing* (1985), Standard 10.4 addresses this issue.

Content validation should be based on a thorough and explicit definition of the content domain of interest. For job selection, classification, and promotion, the characterization of the domain should be based on the job analysis. (p. 60)

The *Uniform Guidelines on Employee Selection Procedures* (1978), which have become the primary standard for adjudication of testing issues in the courts, state in part:

> There should be a job analysis which includes an analysis of the important work behavior(s) required for successful performance and their relative importance and, if the behavior results in work product(s), an analysis of the work product(s). Any job analysis should focus on the work behavior(s) and the tasks associated with them. If work behavior(s) are not observable, the job analysis should identify those aspects of the behavior(s) that can be observed and the observed work products. The work behavior(s) selected for measurement should be critical work behavior(s) and/or important work behavior(s) constituting most of the job. (p. 38302)

The courts' response to inadequate job analysis is typified by the opinion in *Kirkland v. Department of Correctional Services* (cited in Thompson & Thompson, 1982):

> The cornerstone in the construction of a content valid examination is the job analysis. Without such an analysis to single out the critical knowledge, skills and abilities required by the job, their importance relative to each other, and the level of proficiency demanded as to each attribute, a test constructor is aiming in the dark and can only hope to achieve job relatedness by blind luck. (p. 867)

As you can see, there are compelling professional and legal reasons for attending to a careful task analysis. In this chapter we will concentrate on two of the most common techniques used to plan both a course of instruction and its tests: hierarchical analysis of tasks and hierarchical analysis of the levels of learning.

Hierarchical Analysis of Tasks

As you may know, one of the most important steps in a complete course planning process is the creation of the content hierarchy. The hierarchy, in addition to suggesting a logical sequence for instruction, identifies the basic instructional units for the course. Figure 3.1 is an illustration of the relationship of the components that mediate the final goal. Figure 3.2 illustrates a relationship where the three components have been further analyzed and additional

FIGURE 3.1 Hierarchical Relationship of Skills to Course Goal.

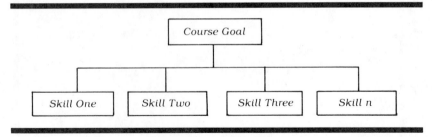

FIGURE 3.2 Extended Hierarchical Analysis.

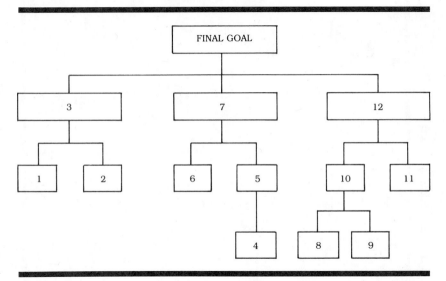

components identified. Finally, Figure 3.3 provides a partial task hierarchy for the skills required by a production-operations manager—someone responsible for the planning, scheduling, and controlling of the production of goods or services.

Matching the Hierarchy to the Type of Test

Let's assume we are planning to teach a unit whose objective is,

- Given a set of data, apply the correct formula to forecast product demand (represented by the "Forecasting" box in Figure 3.3).

FIGURE 3.3   Task Hierarchy of Skills for a Production Manager

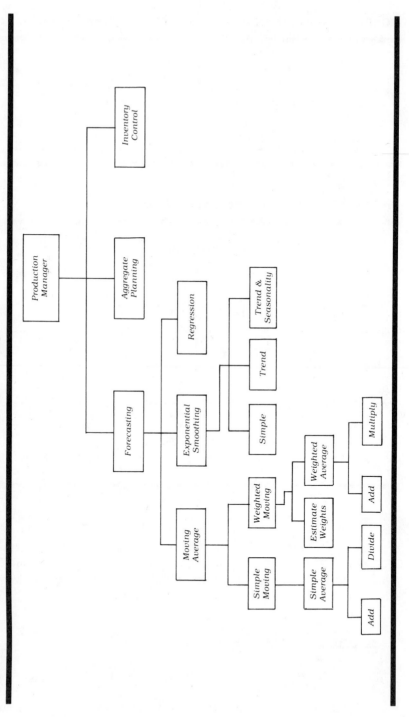

To master this goal, there are three subordinate quantitative skills that need to be mastered:

- Moving average
- Exponential smoothing
- Regression

With the course hierarchy in hand, we can now select objectives to be covered by each of the following five types of tests (described in Chapter 2):

Prerequisite Test. If we want to assure that everyone who enters the course possesses the minimal skills needed to succeed in this unit—skills that will not be covered in the course—we would test at the lowest level of the hierarchy for this unit. In this instance, we would want to be certain that all participants can add, divide, and multiply.

Entry Test. Once we determine that all of our participants possess the basic skills needed to succeed in this unit, we might want to know if they all have the same skill levels, or if some members of the class have already learned some of the material on the job. We might then select objectives that represent skills in each of the boxes two levels below our final objective, i.e., "Simple moving," "Weighted moving," "Simple," "Trend," and "Trend & Seasonality." In a fully individualized course, this test would allow students to bypass instructional units whose objectives they have already mastered.

Diagnostic Test. To determine the exact skill levels of each participant, as well as those areas where they are having problems, we would design a diagnostic test based on all of the skills represented by the boxes below our final objective. Whereas the entry test will typically be used before instruction, the diagnostic test would typically be used in conjunction with the instruction or as part of the post-test process.

Post-Test. At the end of our unit of instruction, we will want to assess student mastery. This type of test would focus on the skills found just below our final objective, i.e., "Moving average," "Exponential smoothing," and "Regression." Test items are drawn from this level because these skills represent the cumulation of all prior skills. Thus the assumption is that if a learner can pass the final objective, then all sub-skills have been mastered.

Equivalency Test. If a manager claims that one of his or her employees does not need the course "because they know it already," we might then ask the employee to take an equivalency test. This test will be composed of items based upon the same objectives as the post-test. (But because having students participate in mastery-based instruction provides a "cushion" for our confidence in their abilities to perform a skill, the equivalency test may need to contain more items in order to establish a similar level of confidence. We will discuss the issue of how test length affects test reliability, and hence validity, in Chapter 4.)

As you plan your test, there are three points pertinent to using content hierarchies that we want you to keep in mind:

- You can't design a good test without good objectives. A fuzzy set of objectives means not only fuzzy instruction, but fuzzy (and therefore invalid) testing.

- Different content hierarchies will result in different specific levels being chosen as the source for items for the different test types. For example, it would be foolish to suggest that items for a prerequisite test will always come from the lowest two levels of the content hierarchy. However, prerequisite items will nearly always come from the lower levels, post-test and equivalency items from the highest levels, and diagnostic and entry test items from nearly all levels.

- You may find yourself using some of the same test items on the different types of tests. That is normal. Just be careful that your learners do not discover that all they have to do is memorize the answers to the pre-test items because the same items will be on the post-test.

USING LEARNING TASK ANALYSIS TO VALIDATE A HIERARCHY

One question that always arises when creating a hierarchy of course objectives is, "Is this hierarchy right?" A valid course hierarchy is especially important in testing, for without this blueprint it is nearly impossible to create an efficient and valid test of any of the five types we have just discussed. A hierarchical sequencing of objectives can only be formally validated through somewhat complex statistical procedures. There is, however, another more workable approach to validating the hierarchy and that is to identify the learning level of each objective and then to see if the levels are in the proper sequence from a learning theory viewpoint.

There are a number of important learning theories that might be used to help validate a hierarchy; we have selected three approaches that have been used in a variety of training settings. These are Bloom's taxonomy (Bloom, 1956), Gagné's learned capabilities (Gagné, 1985), and Merrill's component design theory (Merrill, 1983). If you are familiar with these approaches, you will understand that there is some overlap among the theories; for example, Bloom's "knowledge," Gagné's "verbal information," and Merrill's "remember" levels are similar. However, while each theory makes sense in and of itself, you can't easily (or most likely, usefully) shift back and forth among approaches. Our advice is to pick an approach and stay with it. Then analyze your hierarchies in light of your chosen approach.

BLOOM'S TAXONOMY

Bloom (1956) (and the other psychologists who worked on the taxonomy project) sought to create a system for describing in detail different levels of cognitive functioning so that the precision of testing cognitive performance could be improved. The result of this extensive effort was a classification scheme that breaks cognitive processes down into six types—knowledge, comprehension, application, analysis, synthesis, and evaluation (see Figure 3.4).

The scheme is called a "taxonomy" because each level is subsumed by the next higher level. For example, it is assumed that in order to function at the application level, a person must also be able to function at all levels below application, i.e., comprehension and knowledge. Tasks, objectives, or test items are classified at the high-

FIGURE 3.4 Cognitive Levels of Bloom's Taxonomy

Evaluation
Synthesis
Analysis
Application
Comprehension
Knowledge

est level of cognitive functioning that they require. Therefore, even though analysis level tasks also involve application, comprehension, and knowledge, they are said to be "at the analysis level."

Since its creation, the taxonomy has been widely used to classify the cognitive level of learning objectives and test items. Having a thorough understanding of Bloom's taxonomy is very useful, not only for classifying objectives to validate hierarchies, but also for writing objectives and test items that measure them. The taxonomy provides a test developer with a language for describing different kinds of cognitive operations and therefore provides much guidance in how one might construct test items at these different cognitive levels. Let's look now at the six levels of Bloom's taxonomy in a little more detail.

Knowledge Level. This is the lowest level of the taxonomy and simply indicates the ability to remember content in exactly the same form in which it was presented. Learning objectives that require learners to memorize material—definitions, procedures, formulas, poems, etc.—are said to be written at the knowledge level. Note that the recall of the material is identical to the original presentation in a knowledge level objective. For example, a test item that asks learners to *conduct* a procedure that has been presented to them only verbally is not a knowledge level test item. The knowledge level item would ask learners to *write* the steps in the procedure.

Comprehension Level. Most tasks at this level are of the following four types. Learners restate material in their own words *or* translate information from one form to another *or* apply designated rules *or* recognize previously unseen examples of concepts. Notice that at the comprehension level more than simple rote memorization is required. Learners who must use their own words must do more. Similarly, translation and simple rule application are also forms of representing material using words or symbols other than those used in the original presentation. Obviously, *previously unseen* examples cannot be memorized. If learners were to classify examples previously identified for them, the classification task would be reduced to the knowledge level, because only recall is involved.

Application Level. This task level requires learners to decide what rules are pertinent to a given problem and then to implement the rules to solve the problem. If learners are told what rules to use, the task is reduced to the comprehension level. For example, providing a learner with the formula Area $= \pi r^2$ and a circle's radius and then asking the learner to calculate the area is a comprehension level task. Success requires only recall of the value of π, translating the formula given into the values provided, and multiplication, which is a succession of comprehension level tasks. At the application level, the learner is presented with problems, but not told which rules or formulas to use in solving them. This feature makes application level objectives more complex than comprehension level ones.

Analysis Level. At this level learners are required to break complex situations down into their component parts and figure out how the parts relate to and influence one another. In other words, at the analysis level learners are discovering for themselves what rules explain and govern a given situation. Analysis level tasks ask learners to solve problems after they have been provided extensive scenario descriptions and data, only some of which are pertinent to the solution.

Synthesis Level. Objectives at this level require the creation of totally original material—original products, designs, equipment, etc. Objectives at this level can only be assessed with open-ended kinds of test questions or assignments; multiple-choice or other closed ended questions cannot measure the attainment of synthesis level objectives.

Evaluation Level. The highest level of Bloom's taxonomy, the evaluation level, requires learners to judge the appropriateness or worthwhileness of some object, plan, design, etc., for some purpose. True evaluation level objectives are extremely complex, requiring demonstration of all five lower levels including the synthesis of original criteria for making judgments. Like synthesis level objectives, evaluation level objectives cannot be assessed via closed-ended test questions.

Using Bloom's Taxonomy to Validate a Hierarchy

As we just saw, a basic assumption of Bloom's is that one level subsumes another, i.e., an application level objective would require preceding objectives at the knowledge and comprehension levels. Therefore, when you develop a hierarchy of objectives, you can classify each objective according to Bloom level and then determine if the sequence reflected in your hierarchy matches Bloom's sequence. For example, the hierarchy in Figure 3.5 is correct because lower Bloom levels are taught first. The hierarchy in Figure 3.6 shows a situation where you could expect to have significant instructional

FIGURE 3.5 Correct Hierarchical Sequence

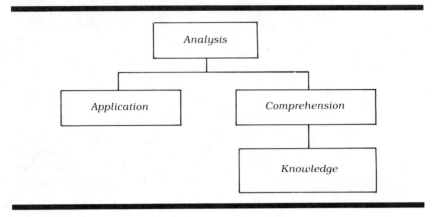

FIGURE 3.6 Incorrect Hierarchical Sequence

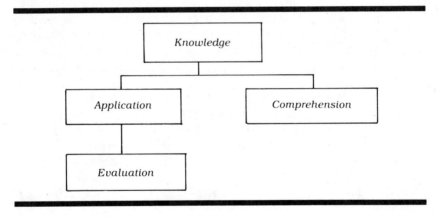

and testing problems, because higher level skills are taught before the prerequisite lower level skills. Figure 3.7 shows how Bloom's levels might be applied to the hierarchy we illustrated in Figure 3.3.

GAGNÉ'S LEARNED CAPABILITIES

Gagné (1985) first divides learning outcomes into five classes of behavior that describe Cognitive skills as well as Motor skills and Attitudes.

FIGURE 3.7 Bloom's Taxonomy Applied to Production Manager Skills Hierarchy

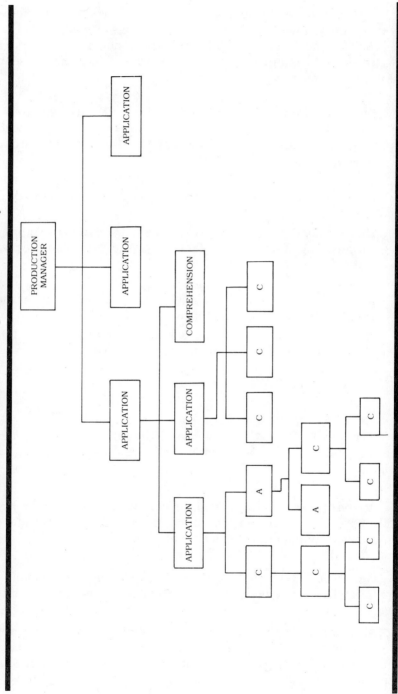

Intellectual Skills. These skills are the ones that enable persons to interact with the environment in terms of symbols and conceptualizations. Intellectual skills are broken down into a hierarchy of five skills. Listed hierarchically from lowest to highest, these five skills are as follows:

- Discriminations—the ability to make different responses to stimuli that differ from each other along one or more physical dimensions; for example, differentiate a "red stoplight" from a "green stoplight."
- Concrete Concepts—the ability to identify a stimulus as a member of a class having common characteristics; for example, a "chair."
- Defined Concepts—the ability to demonstrate understanding of the meaning of some particular class of objects that has no physical referent; for example, "freedom."
- Rules—the ability to formulate a sequence of steps that transforms or identifies a class of objects; for example, "long division transforms a number into its component parts."
- Higher Order Rules (Problem Solving)—complex combinations of simple rules that are created for discovering a solution to a new and previously unencountered situation; for example, diagnosing a nuclear reactor failure.

Cognitive Strategies. These are internal processes by which learners select and modify their ways of attending, learning, remembering, and thinking; for example, practice, paraphrase, note taking, etc.

Verbal Information. Verbal information is the ability to learn specific facts or organized items of information; for example, stating the steps in filling out a service request form.

Motor Skills. Any learned behaviors that lead to movement are motor skills; for example, changing an oil filter, landing a plane.

Attitudes. Attitudes are internal feelings or emotions that affect choices of action; for example, an attitude of professionalism may lead to civility in communications during a performance appraisal.

Using Gagné's Intellectual Skills to Validate a Hierarchy

If you have developed your instructional objectives using Gagné's model, then you can also review your instructional plan through a hierarchical analysis. Since Gagné argues that lower order skills must be learned before higher order ones (as Bloom argues knowledge is prerequisite to comprehension, etc.), any course hierarchy should reflect this prescribed pattern. As a result, any pattern of objectives, when converted to their type of learning, should show that lower level skills, for example, discriminations, must be taught before the higher level skills, for example, concepts. Figure 3.8 illustrates how a set of objectives have been converted to their intellectual skill levels. In this example, the hierarchical structure has been validated by its match with Gagné's prescribed sequence.

MERRILL'S COMPONENT DESIGN THEORY

Merrill (1983) has developed a method of classifying learning tasks as part of an instructional design process called "component design theory." (Originally called component display theory, Merrill has broadened "display" to "design" in recognition of the theory's utility beyond computer display systems.) Merrill's work is very explicit in making one of the most important distinctions in test creation—the difference between memorizing and applying. Component design theory describes two dimensions: the task to be performed and the type of learning.

The Task Dimension. A learner's final performance is divided into three levels: those tasks or relationships the learner must

* remember—repeat exactly as instructed
* use—apply to new situations
* find—create new rules or solutions

Types of Learning. The tasks to be performed can be classified as follows:

* Facts—simple associations between names, objects, symbols, etc.

FIGURE 3.8 Application of Gagné to Hierarchy Validation

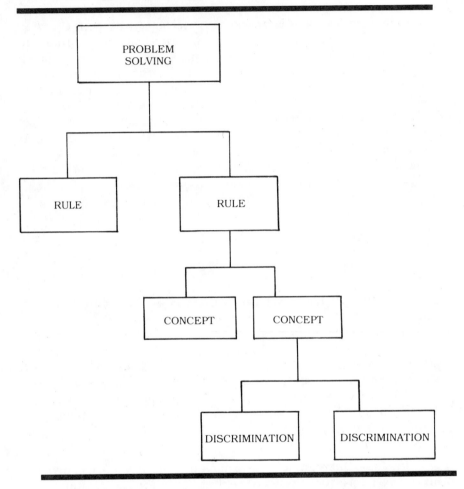

- Concepts—categories or classifications defined by a common set of characteristics
- Procedures—a sequence of specific steps or operations performed on a single type of object
- Principles—explanations or predictions of why things happen based on cause-effect relationships

Thus, in using component design theory, objectives are matched to the type of learning in light of the desired task dimension. For example, an objective such as "The learner will be able to classify a leadership style as Authoritarian, Democratic, or Laissez-Faire" describes concept (type of learning) using (task dimension) behavior. An example test item might be:

- Given a previously unencountered description of a leader's behavior, classify the style as Authoritarian, Democratic, or Laissez-Faire.

Using Component Design Theory to Validate a Hierarchy

As you can probably see, there are stronger similarities between Merrill's work and Gagné's than there are between Merrill's work and Bloom's. This is due to a common foundation in learning theory. And while component design theory would also argue that concepts must be taught before principles, the theory is stronger in pointing out the differences between being asked to repeat something (remember) and applying it (use). Thus, in reviewing your objectives from a component design perspective, one of the most elegant uses of the theory is to simply make sure that no use level task is presented before a remember level task. For example, don't expect sales trainees to practice the six steps of a sales call before they have been taught what the six steps of a sales call are. While this example seems self-evident, the more complex course designs, especially those created by subject matter experts, are more likely to fall into the trap of using before remembering. Figure 3.9 illustrates how a hierarchy might be viewed from Merrill's perspective.

DATA-BASED METHODS FOR HIERARCHY VALIDATION

As we said earlier, more formal, data-based methods exist for validating a hierarchy. These approaches require that data be collected at the end of each instructional module identified by the hierarchy; the data are then analyzed to validate the learning sequence. A real difficulty in this process, however, is deciding whether learner failures are the result of a faulty hierarchy or faulty instruction. In other words, if 100% of the learners pass the first level of instruction and then 50% fail on the second level, is this result due to an improper placement of the instruction in the hierarchy or to poor design of the instructional module?

FIGURE 3.9 Application of Merrill to Hierarchy Validation

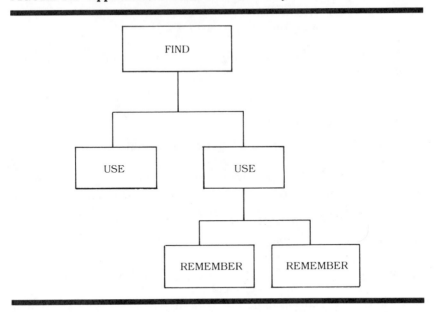

Because data-based analyses of the hierarchy will require well-designed unit tests keyed to each stage of the hierarchy and a relatively large sample of learners, these approaches are usually reserved for highly critical areas of training, for example, nuclear power. The most sophisticated approaches will use Guttman Scalogram analysis to calculate a coefficient of reproducibility (a perfectly scaled hierarchy would have a coefficient of 1.0), as well as Multiple Scalogram analysis to examine the hierarchy for a given optimal sequence of tasks. The underlying theory of these techniques, however, can be demonstrated and used by almost any course designer through the simple analysis of post-test data percentages.

In Figure 3.10, the post-test scores are shown for each level in a simple hierarchy. As you can see, the pattern of success is about what we would expect for a valid hierarchy with good (not perfect) instruction. There are high levels of performance at the bottom, indicating mastery of the content, with some attrition in performance as the task becomes more complex at the top of the hierarchy.

In Figure 3.11, there was success at the lowest levels, but the pattern of performance in the higher levels is confusing. The "40-75-25" pattern could indicate a number of problems, but a first reaction might be that the unit where learners scored 40% is unnecessary, as many scored much higher at the next level (75%). The

FIGURE 3.10 Analysis of Post-test Scores to Validate a Hierarchy, Example of a Valid Hierarchy

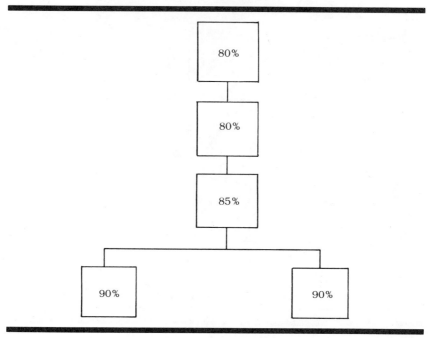

drop in performance at the top of the hierarchy may also be due to a needed, but missing, unit of instruction, or to weak instruction in the unit itself.

WHO KILLED COCK ROBIN?

In the end, it is important to remember that while a hierarchy provides the test designer with a blueprint for making efficient testing decisions, the hierarchy itself may not be valid for the job skills to be assessed, even though it is internally logical and correctly sequenced. In other words, a hierarchy could be analyzed through a task analysis, a learning analysis, and then verified through empirical techniques invoking the most sophisticated Guttman scalogram analysis to determine that the sequencing of tasks is correct—and remain invalid because the wrong tasks were analyzed. In *Fire Fighters Institute for Racial Equality v. City of St. Louis* (cited in Barrett,

FIGURE 3.11 Analysis of Post-test Scores to Validate a Hierarchy, Example of an Invalid Hierarchy.

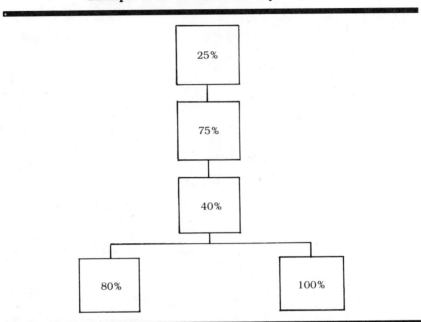

1981), the court examined the content validity of an exam for the position of fire captain and noted the following:

> Constructing a content valid exam and proof of its validity requires as a first step a thorough analysis of the job to be performed. . . . It is in fact the fatal flaw in the validation study that the test . . . devised did not reflect [the] findings in the job analysis. The captain's exam admittedly failed to test the one *major* job attribute that separates a firefighter from a fire captain, that of supervisory ability. . . . The job analysis here may have appeared impressive in relation to those challenged in other cases, but a good analysis in any situation is of little use when the examination fails to reflect what is found in the job analysis. The test is not *content* valid. . . . Here, where the exam failed to test a job component comprising over 40 percent of the employee's time, the inference of discrimination has not been rebutted with a finding of the exam's "job relatedness." (p. 591)

Barrett's article, "Is the Test Content-Valid: Or, Who Killed Cock Robin?," summarizes the courts' attitudes toward job relatedness and test items. The article illustrates the role of job relatedness with an item the courts found to be content-invalid for assessing the skills required of elementary school principals:

> Of the following characters in the nursery rhyme, "The Burial of Poor Cock Robin," the one who kills Cock Robin is the
> 1. Lark
> 2. Thrush
> 3. Bull
> 4. Sparrow (p. 589)

Possibly a job-related item, but not valid for the task to be assessed.

4.

Item Creation for Paper-and-Pencil Tests

THE ROLE OF OBJECTIVES IN TEST ITEM WRITING
CLASSIFICATION SCHEMES FOR OBJECTIVES
TYPES OF TEST ITEMS
GUIDELINES FOR WRITING TEST ITEMS
HOW MANY ITEMS SHOULD BE ON A TEST?

THE ROLE OF OBJECTIVES IN TEST ITEM WRITING

It would be difficult to overstate the usefulness of good instructional objectives in the creation of sound tests. Most instructional designers are aware of how important objectives are to the creation of instruction; many are less familiar with the role of objectives in testing. It is not the purpose of this book to treat comprehensively the techniques for writing instructional objectives. However, we do want to focus briefly on the critical role objectives play in testing and the components of well-written objectives most essential for item writing.

Instructional objectives serve three fundamental purposes for criterion-referenced test developers:

• Objectives ensure that the test covers those learner outcomes important for the purposes that the test must serve. Remember that there are several different types of tests (see Chapter 2) and that the content for these tests is derived by task analysis procedures that order objectives hierarchically. Matching test items to the appropriate course objectives within these hierarchies guarantees that all essential content is assessed.

- Objectives increase the accuracy with which cognitive processes in particular can be assessed. A well-written objective is a blueprint for the creation of test items that will assess the specific competency described by the objective. In this way, objectives make it much easier for test writers to create so-called parallel test items, i.e., different test items that assess the same objective. Parallel test items are essential for the construction of reliable tests, as we shall see later in this chapter, as well as for the creation of equivalency tests and different forms of any given test. Hence objectives are essential to the construction, maintenance, and security of an organizational testing process.

- The size of the domain covered by the objectives and the homogeneity of the objectives the test is designed to assess are important factors in determining how many items will need to be included on the test. These characteristics of objectives are discussed later in this chapter in the section on determining test length.

Characteristics of Good Objectives

Of course, not all objectives are equally well written. Numerous authors have provided course developers with advice about how to write objectives. Most authorities agree, however, that good objectives have four parts:

- Who the learner is
- What behavior or competency the learner will perform
- Under what conditions the learner will perform the competency
- To what standard of correctness the learner will perform the competency

We will examine briefly the latter three components of the objective—the behavior, the conditions, and the standard. An understanding of the learner is, of course, an extremely important part of the instructional design process, but for our purposes we are assuming that this learner analysis has already been completed and that, as test designers, we know the pertinent characteristics of those for whom our learning objectives and our tests are intended.

The Behavior Component. It is essential that the competency that the learner is to perform be described in observable, measurable terms, hence the term "behavioral objective" used to describe the most useful statements of learner outcomes. When writing objectives, choose the most precise verb you can to state what the learner will be able to do. For example, the words "list," "categorize," "draw," and "evaluate" are better than "understand," "appreciate," "know," and "really understand" as verbs in behavioral objectives. The more descriptive the verb in an objective, the easier it will be to write test items that accurately assess the objective.

The Conditions Component. If well written, this part of the objective provides useful information to test writers, since the test essentially presents learners with a series of conditions under which they must demonstrate their achievement of the instructional objectives. Unfortunately, the conditions component of an objective is frequently omitted by designers who do not realize how critical it is for clearly communicating the intent of the objective. Changing the conditions under which a behavior is to be performed can dramatically alter the difficulty and nature of the competency assessed.

For example, the behavior "assemble the milk shake machine" is significantly altered depending upon whether the corresponding condition is "given the unassembled parts and the repair manual" or simply "given the unassembled parts." The behavior with the latter condition can be expected to be significantly more difficult than with the former, and in fact the very nature of the intended competency specified by the objective changes depending upon which condition is used. Under the former condition the objective describes skills in reading and using a repair manual, whereas under the latter condition the objective specifies mechanical analysis skills.

The Standards Component. Complete objectives include a statement of how well the learner must perform the indicated behavior. This component, however, is probably the most difficult component to write. It frequently takes the form "with 90% accuracy" or "correctly 80% of the time." It is helpful to realize that all standard statements need not be in the form of percentages. In fact, many competencies do not lend themselves to percentage standards at all. Other forms of standards are in terms of

- number of allowable errors
- time limits
- expert judgments
- negative consequences avoided, for example, "remove pizza from oven to boxing counter *without burning the fingers*"

If available, the standards component can be useful to test writers in setting the cutoff score for mastery of a criterion-referenced test. However, as we shall see in Chapter 9, the setting of the cutoff score is a somewhat complex procedure. Test writers need to be careful of objectives with hastily written, ill-thought out standards statements. Test developers cannot afford simply to adopt the standards dashed off by course designers who do not understand the enormous significance the cutoff score has to the test's effectiveness and its ability to withstand legal challenge.

CLASSIFICATION SCHEMES FOR OBJECTIVES

As we have seen, good objectives are an essential precursor to sound testing systems. Translating objectives into rating scales for performance tests (Chapter 5) is usually easier than translating objectives into test items for paper-and-pencil tests. One strategy that can be helpful in this regard is to first classify the objectives according to the type of cognitive behavior each requires. Classifying objectives by cognitive skill assists item writers in

- choosing which item type—mutiple-choice, essay, etc.—will most accurately and efficiently assess the objective, and
- deciding what the text of each item will be.

Several different classifications of cognitive behavior have been developed over the years. We discussed three of them, the systems written by Bloom, Gagné, and Merrill, in Chapter 3. In this chapter, however, we will concentrate on specific item types in relation only to Bloom's classification system for a specific reason. Gagné and Merrill based their classifications on learning theory approaches. Their systems, therefore, are very useful for instructional designers because they provide instructional prescriptions for each type of learning task. In contrast to the learning theory approaches of Gagné or Merrill, Bloom and his colleagues developed their system through an intensive content analysis of thousands of instructor-created test items. As a result, Bloom's Taxonomy provides a particularly comfortable fit with and support to cognitive assessment. You may find the other schemes helpful for test writing as well as instructional design, and all three do share some commonalities, but Bloom's Taxonomy probably provides more refined guidance to test item construction.

Bloom's Cognitive Classifications

In Chapter 3 we described Bloom's Taxonomy of cognitive objectives in some detail. As you will recall, the Taxonomy was created to improve precision in the testing of cognitive processes. The classification scheme consists of six levels with each given level subsuming all levels beneath it, as follows:

- Evaluation
- Synthesis
- Analysis
- Application
- Comprehension
- Knowledge

Each of these cognitive levels is described in great detail in the book that originally presented the Taxonomy, *Taxonomy of Educational Objectives. Handbook I: Cognitive Domain* (Bloom, 1956). Understanding the nature of the cognitive performance to be assessed is a good first step to being able to write an appropriate test item. In the *Handbook* the description of each cognitive level is accompanied by many examples of test items that assess that particular cognitive behavior. If a test writer can correctly identify the Bloom level of an instructional objective, a wealth of ideas about how to measure the objective become available.

Another important result of understanding Bloom's Taxonomy is an increased awareness of all the cognitive behaviors beyond simply remembering, i.e., beyond the knowledge level. Most of the tests we take in school at all grade levels and even at the college level are composed of knowledge level questions. This circumstance is not difficult to explain, since knowledge level items are by far the easiest to write. However, developing tests that truly reflect on-the-job performance requires the ability to distinguish among different cognitive behaviors and skill in writing items at the higher cognitive levels, particularly the comprehension, application, and analysis levels.

PRACTICE

Here are six instructional objectives. Indicate at what Bloom cognitive level you think each objective is. Remember that objectives are classified at the highest level of cognitive functioning that they require.

1. Given an historical account of a working task force, identify the major characteristics of the group's functioning and describe the causal relations among these characteristics that explain the group's behavior.

   _____

2. Shown a videotape of a business meeting, use principles of group behavior to predict the likely outcomes of the meeting.

   _____

3. Given access to a job description, a subject matter expert, and other supporting documentation regarding job responsibilties and employee characteristics, design an appropriate course of instruction to train an employee to perform the job.

   _____

4. Given an oral description of a procedure, depict the procedure as a flowchart.

   _____

5. List the criteria presented in class for judging the effectiveness of an oral presentation.

   _____

6. Given a marketing plan for a new product, a description of the product, and access to the product's designers, determine the likely effectiveness of each major stage of the plan.

   _____

FEEDBACK

1. Analysis
2. Application
3. Synthesis
4. Comprehension
5. Knowledge
6. Evaluation

TYPES OF TEST ITEMS

There are six types of test items commonly used in paper-and-pencil tests:

- True/False
- Matching
- Multiple-choice
- Fill-in
- Short answer
- Essay

Of these six, multiple-choice is the preferred item type for most paper-and-pencil tests. It has the advantage of being able to assess most of Bloom's cognitive levels and yet can be reliably scored by hand or by machine. Therefore, throughout our discussion of item types, we will frequently make comparisons between a given item type and multiple-choice. For each of these six item formats, we present

- a description of the item type and the kind of content for which the format is best suited
- the Bloom levels assessable by the item type
- the major advantages and disadvantages of using the item type
- a summary of the guidelines for writing each item type correctly

True/False Items

Description. The true/false item presents the test-taker with a statement that he or she must indicate is either true or false. This type of item is a sensible choice for "naturally dichotomous" content, i.e., content that presents the learner with only two plausible choices. For example, assume your objective requires that, given blood composition data, learners will classify the blood as that of a male or a female. You might construct a true/false question asserting that a given blood composition is male, to which the test-taker would respond "true" or "false." Content that is not naturally dichotomous is usually best assessed using the multiple-choice format, because true/false questions have some distinct limitations, which will be discussed below.

Bloom Levels. True/false items can assess the knowledge, comprehension, and application levels. Unfortunately, however, they are most often used to assess only the knowledge level.

Advantages. The primary advantage of true/false items is that they are typically easier to write than other types of closed-ended questions, i.e., matching or multiple-choice. However, the reputed ease of construction is partly because most of these items are written at the knowledge level; it requires more thought to write true/false items at higher cognitive levels. Their other advantages are that, like all closed-ended questions, they are easily and reliably scored, and test-taker responses can be submitted to statistical item analysis that can be used to improve the quality of the test. (These item analysis procedures are discussed in Chapter 8.)

Disadvantages. The biggest disadvantage of true/false items is that test-takers have a fifty-fifty chance of getting the items correct simply by guessing. However, if the content that the true/false item covers is truly dichotomous, a multiple-choice item with more than two choices would be very difficult to write anyway. Note that multiple-choice items with only two choices also allow test-takers to guess correctly half of the time. Before writing true/false items, always examine the content and instructional objectives carefully to be sure that they are not more appropriately addressed by multiple-choice items. The key to using true/false items effectively is to use them only when the content is naturally dichotomous and to write true/false items that require more than mere memorization of content.

Matching Items

Description. Matching items present test-takers with two lists of words or phrases and ask the test-taker to match each word or phrase on one list (hereafter referred to as the "A" list) to a word or phrase on the other (the "B" list). These items should be used only to assess understanding of homogeneous content, for example, types of wire, types of clouds, types of switches, etc. Matching items most frequently take the form of a list of words to be matched with a list of definitions.

Bloom Levels. Matching items can assess the knowledge and comprehension levels. However, like true/false items, we very rarely see them written beyond the knowledge level.

Advantages. Matching items are relatively easy to write. Note, however, that one reason for this feature is that they do not assess beyond the comprehension level. Matching items can be scored quickly and objectively by hand and frequently also by machine. Responses to matching questions can be submitted to statistical item analysis procedures.

Disadvantages. Matching items are limited to the two lowest levels of Bloom's Taxonomy. Another disadvantage is that if these items are constructed using heterogeneous content, i.e., if the words or phrases appearing on the "A" list are essentially unrelated to one another, matching items become extremely easy. For example, a list that contains a type of wire, a type of cloud, a type of switch, etc., will be easier to match to a corresponding "B" list than will a list that contains only names of different types of wire. Another difficulty with matching items results from test writers including equal numbers of entries in both lists or allowing items from the "B" list to be used only once. Under these circumstances test-takers can use the process of elimination to figure out cues to the correct matches.

Multiple-Choice Items

Description. The multiple-choice item presents test-takers with a question (technically called a "stem") and then asks them to choose from among a series of alternative answers (a single correct answer and several distractors). Sometimes the question takes the form of an incomplete sentence followed by a series of alternative completions from which the test-taker is to choose one. Sometimes the stem is a relatively complex scenario containing several pieces of information ending in a question. Dichotomous content can be assessed using multiple-choice questions with two optional answers; thus most true/false items can be converted to the multiple-choice format.

Bloom Levels. Multiple-choice questions can assess all Bloom levels except the two highest ones, synthesis and evaluation. The reason that these two levels are beyond the multiple-choice format is that they require totally original responses on the part of the test-taker. Since multiple-choice questions are closed-ended, i.e., the correct answer appears before the test-taker who must recognize it, the test-taker's response is necessarily not original. However, multiple-choice allows assessment of more Bloom levels than any other closed-ended question format.

EXAMPLES OF MULTIPLE-CHOICE ITEMS AT DIFFERENT
BLOOM LEVELS

Here are four multiple-choice items, one written at each of the
four Bloom levels assessable by items of this format.

Knowledge Level

According to Gagné, the association of an already available response with a new stimulus is called
a. instrumental conditioning

b. learning

c. signal learning

d. front-end analysis

Discussion. The answer is "c." The item is knowledge level because it simply asks the student to remember the definition of
signal learning.

Comprehension Level

According to the definition of "refugee" used by the United Na-
tions, what group below would **not** qualify as refugees?
a. Vietnamese boat people in Indonesia

b. Ugandan Christians who fled to Tanzania during Idi Amin's
rule

c. Cambodian followers of Pol Pot who fled to the northwestern
mountains of Cambodia when the Vietnamese invaded Cambodia in December, 1978

d. Palestinians now living in Lebanon

e. Jews who came to the United States from Europe in the 1930s

Discussion. The answer is "e." This item is written at the comprehension level because the test-taker has been asked to translate information from one form to another. In this item the test-
taker must translate from one level of abstraction to another, i.e.,
the student has been asked to recognize a specific non-example
(the word "not" is emphasized) of the more abstract concept "refugee." For an item to be at this level, the test-taker must be able
to do more than simply remember the correct answer. This item
would not be a comprehension level item, but a knowledge level
item if the students had previously been shown these examples
of refugees or told that Jews who came to the United States during
the 1930s were not refugees.

Many times a test designer cannot tell whether an item is at
the knowledge level or any other level without knowing the content of the students' previous instruction. Any item that has been
previously encountered will always be a knowledge level item
regardless of the test designer's intent.

Application

Amy, three and a half years old, spills her milk at the table. According to current principles of child development, her parents should

a. tell her she shouldn't waste milk, and refuse to let her have any more at the meal.

b. wipe up the milk in silence and serve her again.

c. have Amy get the cloth to help wipe it up. Serve her again.

d. dismiss Amy from the table.

e. tell Amy that she is a baby when she spills, and that the next time she will be served milk from a bottle.

Discussion. The answer is "c." This is an application item emphasizing the application of child development principles in a new context. It is assumed that the students have never before been confronted with Amy's milk-spilling behavior. They are being asked to apply a principle of child development having to do with encouraging children to accept responsibility for the consequences of their behavior. In choosing the correct answer, the test-taker must consider the situation, decide how it is similar to the content in which he or she learned relevant principles, and then apply the correct principles to get the right answer.

Analysis

Susan, a student in Mr. Stepp's statistics class, asks Mr. Stepp what her average score is for the three exams he has given the class. Breaking with his usual manner of reporting scores, he replies that her average is +1.7. Which of the following assumptions about the students' scores on these tests is most plausible?

a. The standard deviations of scores on all three tests were similar.

b. None of the tests produced extremely skewed distributions.

c. All of the students did poorly on at least one of the tests.

d. The correlations between the three sets of test scores were restricted.

Discussion. The answer is "d." This is an analysis item because it requires the test-taker to recognize unstated assumptions. In this item the test-taker must first recognize that Mr. Stepp has converted the students' raw test scores to standard Z scores, a type of score based on the student's position relative to the other test-takers. The test-taker must then remember that this conversion is usually made before averaging the scores from tests, which have resulted in very unlike distributions, or in a lack of variation in the scores on one or more of the tests.

   It is this assumption of lack of variation in the scores that provides the clue to the correct answer. The only choice that makes a true statement about test scores generally lacking in variation is "d." The correlations between tests when students tend to score alike—when there is a lack of variation in their scores—will be low, or, more technically speaking, "restricted." Hence "d" is the correct answer. Assumption "a" is wrong because similar standard deviations on tests indicate like distributions. Assumption "b" is wrong because the absence of extremely skewed scores increases the likelihood that adequate variation was present in the scores. Assumption "c" is wrong because all of the students could have done well on one of the tests and the variation would still be lacking. In other words, "c" is an assumption too specific to be warranted by the facts given.

Note the differences between application and analysis. In application, the emphasis is on remembering and bringing to bear upon given material the appropriate principles. In analysis, the emphasis is on breaking complex material down into its component parts and detecting the relationships of the parts to one another. Remember, analysis does involve application, as it does comprehension and knowledge. Such is the hierarchical nature of Bloom's Taxonomy.

Advantages. Multiple-choice is the most flexible of all closed-ended item formats. Multiple-choice items can assess any kind of content at a variety of Bloom levels. Because the test-taker must choose among several optional answers, the probability of simply guessing the correct answer is lower than with true/false items. Furthermore, multiple-choice items are ideal for diagnostic testing. In other words, the distractors can target those learners who have specific problems; knowing what wrong answers test-takers chose can be important and useful information for instructors and course designers. In addition, multiple-choice questions are quickly and reliably scored either by hand or by machine and are ideally suited to statistical item analysis procedures that can lead to improved test quality (see Chapter 8).

Disadvantages. The major disadvantage of multiple-choice questions is that they are difficult and time consuming to write. Most testing authorities agree that well-written multiple-choice questions are usually worth the effort, especially if they can be used repeatedly with a large number of test-takers. An additional weakness is that multiple-choice questions cannot assess objectives that require test-takers to recall information unassisted, since the correct answer does appear before the test-taker among the options. Their only other disadvantage is their inability to assess directly the synthesis and evaluation cognitive levels.

Fill-In Items

Description.   Unlike the first three item formats discussed, fill-in items are open-ended, i.e., the answer does not appear before the test-taker. Rather, the fill-in item is a question or an incomplete statement followed by a blank line upon which the test-taker writes the answer to the question or completes the sentence. Therefore, fill-in questions should be used when the instructional objective requires that the test-taker recall or create the correct answer rather than simply recognize it. Objectives that require the correct spelling of terms, for example, require fill-in items. Fill-in items are limited to those questions that can be answered in a word or short phrase; short answer and essay questions require much longer responses.

Bloom Levels.   Fill-in items can assess the knowledge, comprehension, and application levels. They are written most often, however, at the knowledge level.

Advantages.   Fill-in items are typically easy to write. They are essential for assessing recall as opposed to recognition of information.

Disadvantages.   There are two major disadvantages of fill-in items. One is that they are suitable only for questions that can be answered with a word or short phrase. This characteristic typically limits the sophistication of the content that can be assessed with fill-in items. The second major disadvantage is that, like all open-ended questions, fill-in items present scoring problems. Because test-takers are free to write any answer they choose, sometimes there can be a debate over the correctness of a given answer. Test-takers are marvelously unpredictable when it comes to concocting an unanticipated answer to an open-ended question. Unlike the scoring of closed-ended questions, the scoring of all open-ended questions requires judgment calls on the part of the scorer.

Short Answer Items

Description.   These items are open-ended questions requiring responses from test-takers of one page or less in length. Short answer questions require responses longer than those for fill-in items and shorter than those for essay questions. Short answer questions are recommended when the objective to be assessed requires that the test-taker recall information unassisted (rather than recognize information) or create original responses of relatively short length.

Bloom Levels. Short answer questions can be used to assess all Bloom levels except possibly the highest one, evaluation; most responses to evaluation questions would necessarily be somewhat longer.

Advantages. The major advantage of short answer questions is that they are able to elicit original responses from test-takers. For some objectives at the higher Bloom levels, only short answer and essay questions are appropriate. Lower level short answer questions are typically easier to write than multiple-choice questions covering the same content. It is important to remember, however, that changing the format of a question can significantly alter the cognitive skills assessed. Short answer items are best reserved for those objectives that cannot be assessed using closed-ended questions.

Disadvantages. The disadvantages of short answer questions are, unfortunately, extremely serious ones. Most notably, short answer questions are very difficult to score reliably. The evaluation of short answer responses and essays are notoriously prone to error—resulting from halo effects, the placement of a given test in the scoring sequence, scorer fatigue, and, especially, quality of handwriting. In addition to being unreliable, the scoring of short answer responses is time consuming. Short answer questions also require far more time to answer than multiple-choice questions, thus sometimes limiting severely the content that can be covered by the test.

Essay Items

Description. Essay items are open-ended test questions requiring a response longer than a page in length. They are recommended for objectives that require original, lengthy responses from test-takers. Essay items are also recommended for the assessment of writing skills.

Bloom Levels. Essay questions can be used to assess all levels of Bloom's Taxonomy. They are the only item type with this capability, and the only item type that can truly assess the evaluation level.

Advantages. The essay question's major advantage is its capacity to assess the highest cognitive levels. Essay questions that assess the lower levels are usually not difficult or time consuming to construct. Those that assess the higher levels can be very difficult to write, requiring the provision of a great deal of stimulus material to which the test-taker responds in the essay.

Disadvantages. The disadvantages of the essay item are identical in nature to those of the short answer item; however, these problems are aggravated by the additional length of the responses. Essay questions are even more difficult to score reliably, take even more time to score, and use up even more testing time than do short answer questions. For these reasons, essay items are to be avoided if at all possible. Use essay questions only when the cognitive level of the objective requires it.

GUIDELINES FOR WRITING TEST ITEMS

This section presents a summary of guidelines for writing each of the six types of test items—true/false, matching, multiple-choice, fill-in, short answer, and essay.

Guidelines for Writing True/False Items

- Use true/false items in situations where there are only two likely alternative answers, i.e., when the content covered by the question is dichotomous.
- Include only one major idea in each item.
- Make sure that the statement can be judged reasonably true or false.
- Statements should be as short and simply stated as possible.
- Avoid negatives, especially double negatives; highlight negative words (for example, *not, no, none*) if they are essential.
- Any statement of opinion should be attributed to its source.
- Randomly distribute the true and false statements.
- Avoid specific determiners (for example, always, never) in the statements.

Guidelines for Writing Matching Items

- Include only homogeneous, closely related content in the lists to be matched.

- Keep the lists of responses short—five to fifteen entries.
- The response list should be arranged in some logical order, for example, chronologically or alphabetically.
- Directions should clearly indicate the basis on which the entries are to be matched.
- Directions should indicate how often a response can be used; responses should be used more than once to reduce cuing due to the process of elimination.
- Use a larger number of responses than entries to be matched in order to reduce process of elimination cuing.
- Place the list of entries to be matched and the list of responses on the same page.

Guidelines for Writing Multiple-Choice Items

Guidelines for Writing the Stem.

- The stem should be written using the simplest and clearest language possible to avoid making the test a measure of reading ability.
- Place as much wording as possible in the stem, rather than in the alternative answers; avoid redundant wording in the alternatives.
- If possible, state the stem in a positive form.
- Highlight negative words (no, not, none) if they are essential.

Guidelines for Writing the Distractors.

- Provide three or four alternative answers including the correct response.
- Make certain you can defend the intended correct answer as clearly the best alternative.
- Make all alternatives grammatically consistent with the stem of the item to avoid cuing the correct answer.
- Vary randomly the position of the correct answer.
- Vary the relative length of the correct answer; don't allow the correct answer to be consistently longer (or shorter) than the distractors.
- Avoid specific determiners (all, always, never) in distractors.

- Use incorrect paraphrases as distractors.
- Use familiar-looking or verbatim statements that are incorrect answers to the question as distractors.
- Use true statements that do not answer the question as distractors.
- Use common errors that students make in developing distractors; anticipate the options that will appeal to the unprepared test-taker.
- Use irrelevant technical jargon in distractors.
- Avoid the use of "All of the above" as an alternative; test-takers who recognize two choices as correct will realize that the answer must be "All of the above" without even considering the fourth or fifth alternatives.
- Use "None of the above" with caution; make sure it is the correct answer about one third to one fourth of the times it appears.
- Avoid alternatives of the type "Both a and b are correct," or "a, b, and c but not d are correct," etc.; such items tend to test a specific ability called "syllogistic reasoning," as well as the content pertinent to the item.
- Items with different numbers of options can appear on the same test.
- If there is a logical order to options, use it in listing them; for example, if the options are numbers, list them in ascending or descending order.
- Check the items to ensure that the options or answer to one item do not cue test-takers to the correct answers of other items.

Guidelines for Writing Fill-In Items

- State the item so that only a single, brief answer is likely.
- Use direct questions as much as possible, rather than incomplete statements, as a format.
- If you must use incomplete statements, place the blank at the end of the statement, if possible.
- Provide adequate space for the test-taker to write the correct answer.
- Keep all blank lines of equal length to avoid cues to the correct answers.

- For numerical answers, indicate the degree of precision required (for example, "to the nearest tenth") and the units in which the answer is to be recorded (for example, "in pounds").

Guidelines for Writing Short Answer Items

- State the question as clearly and succinctly as possible.
- Be sure that the question can truly be answered in only a few sentences rather than requiring an essay.
- Provide guidance regarding the length of response anticipated (for example, "in 150–200 words . . .").
- Provide adequate space for the test-taker to write the response.
- Indicate whether spelling, punctuation, grammar, word usage, etc., will be considered in scoring the response.

Guidelines for Writing Essay Questions

- State the question as clearly and succinctly as possible; present a well-focused task to the test-taker.
- Provide guidance regarding the length of response anticipated (for example, "in 5–6 pages . . .").
- Provide estimates of the approximate time to be devoted to each essay question.
- Provide sufficient space for the test-taker to write the essay.
- Indicate whether spelling, punctuation, grammar, word usage, etc., will be considered in scoring the essay.
- Indicate whether organization, transitions, and other structural characteristics will be considered in scoring the essay.

WRITING DIFFERENT TYPES OF TEST ITEMS

We have discussed six types of items. Here is a summary of the six types of items and their most appropriate applications:

- True/False—naturally dichotomous content for knowledge, comprehension, and application levels

- Matching—homogenous material for knowledge and comprehension levels
- Multiple-Choice—most flexible item for any cognitive level except synthesis and evaluation
- Fill-In—tests recall rather than recognition at the knowledge, comprehension, and application levels for a word or phrase
- Short Answer—tests recall or original creation of relatively short length for all levels but evaluation
- Essay—tests writing skill and original creation of some length for any cognitive level

PRACTICE

For each of the following objectives, write a test item of the indicated type:

1. Write a true/false item for the following objective: Given the name of a previously unclassified plant or animal, classify the name as that of a plant or an animal.

   _____

   _____

   _____

2. Write a matching item for the following objective: Given a list of characteristics that distinguish one breed of feline from another and a list of feline breeds, match each breed with its distinguishing characteristics.

   _____

   _____

   _____

   _____

3. Write a multiple-choice item for the following objective: Given consumption, government spending, investment, and net foreign investment, calculate the Gross National Product.

   _____

   _____

   _____

   _____

4. Write a fill-in item for the following objective: Given a brief description of the distinguishing attributes of a plant, write the correct spelling of the plant's common name.

_____

_____

_____

5. Write a short answer item for the following objective: Given a description of a family seeking to purchase a pet, provide the rationale for the selection of an appropriate animal.

_____

_____

_____

_____

6. Write an essay item for the following objective: Given the professional position of a car buyer, personal data, and the model purchased, evaluate the appropriateness of the purchase including the specification of the criteria used in the evaluation.

_____

_____

_____

_____

FEEDBACK

1. T　F　　Staphylococcus is a plant.

2. On the left below is a list of feline breeds. On the right are feline characteristics that distinguish one breed from another. Match each breed with that characteristic that serves as one of its primary indicators. Record the corresponding characterstic letter in the blank to the left of each breed. Characteristics may be used more than once.

_____ 1. Siamese　　　　a.　blue eyes

_____ 2. Abyssinian　　　b.　mane

_____ 3. Persian   c. rounded ears

_____ 4. Himalayan   d. almond eyes

_____ 5. Burmese   e. rough coat

                                       f. tailless

3. If consumption is $600 billion, investment is $100 billion, net foreign investment is $8 billion, and government spending is $250 billion, what is the Gross National Product?

   a. $758 billion

   b. $858 billion

   c. $950 billion

   d. $958 billion

4. What is the name of a very dangerous spherical bacteria occurring in irregular clusters?

   _____

5. The Larsens want to acquire a cat. Their annual income is $15,000. They live in a small apartment. They have two children aged five and seven; the children will be taking care of the cat. Their living room furniture is navy blue.
   Select a feline breed for the Larsens and in 100 to 150 words explain why you think your choice is appropriate.

6. The President of the United States has decided to purchase a Chevrolet Chevette. Write a well-organized essay of not more than 1500 words evaluating the appropriateness of this automobile for the President. Be sure to state clearly the criteria you used in making your evaluation.

HOW MANY ITEMS SHOULD BE ON A TEST?

It is always useful to create more test items than you think you will need in case some of the items appear to be flawed after you review the item analysis data (see Chapter 8). At some point in the test planning process, however, developers need to decide how many items will appear on the test. This is a question that, unfortunately, does not have a simple, numerical answer. It is an extremely important question, however, because the length of the test has a direct relationship to the test's reliability and therefore to its validity as well. The question of test length turns on at least four factors:

- The criticality of the mastery decisions made on the basis of the test
- The resources (time and money) available for testing
- The domain size described by the objectives to be assessed
- The homogeneity or relatedness of the objectives to be assessed

This section on test length begins with a brief discussion of test reliability and why more items are always better for test reliability than fewer items. Each of the four factors listed above is then discussed in terms of its influence on the decision of test length. The section closes with a brief look at the advice resulting from research into test length and a summary that integrates the factors that impinge on the test length issue.

Test Reliability and Test Length

As you may recall from Chapter 1, test reliability for paper-and-pencil tests has to do with the test's consistency in results over time. Consistency for criterion-referenced tests means consistency in classifying test-takers as masters or nonmasters. In other words, if an employee takes the test twice during the space of two weeks, a reliable test will classify the employee the same way both times. The longer the test, the more consistent it is likely to be.

To understand why longer tests are more reliable than shorter tests, consider this analogy. Pretend you are standing in front of a large opaque, black jar that is filled with three different colors of jelly beans—some red, some green, and some yellow—all mixed together in the jar. Your task is to estimate approximately what proportion of the beans is of each color. You are allowed to sample the beans, i.e., you are allowed to take some out and examine them in order to help you make this decision. You reach in and take out a sample of three beans; two are red and one green. This result might lead you to think that there must be more red beans in the jar than any other color, and probably more green than yellow. That would be a plausible guess, but as you might imagine you would not have a great deal of confidence in your ability to specify the proportions of colors present in the entire jar based on so small a sample. So you draw another sample and another. Eventually you have drawn out about three fourths of the beans. Based upon a sample of this size, you would be fairly confident in estimating the proportions of beans of each color in the jar.

Test items are like your samples of jelly beans; they are your opportunity to sample what the test-taker can do. As in the jelly bean example above, the more samples you draw, i.e., the more items you include, the more accurate your picture of the competence of the test-taker will be. Hence your decisions regarding the mastery or nonmastery status of your test-taker will be more consistent when based upon a longer test than upon a shorter test.

While the generalization that "more is better" is valid, clearly some constraints must be placed on test length. Here is where the four factors mentioned above become important considerations.

Criticality of Decisions and Test Length

We know that test reliability is a function partly of test length. Therefore, when trying to decide how many items to put on a test, it makes sense to ask the question, "How reliable does the test have to be?" Sometimes errors in master/nonmaster classification of test-takers can be tolerated. It is very useful to do a systematic analysis of the consequences of both types of classification errors resulting from unreliable criterion-referenced tests.

Ask yourself and others who are knowledgeable about the responsibilities of the target test-takers, "What are the costs to this organization of erroneously classifying a nonmaster as a master?"—sometimes called an "error of acceptance" or a "false positive error." Undeserved bonuses? Poor work performance? Law suits from clients? Deaths? And "What are the costs to the organization of erroneously classifying a master as a nonmaster?"—sometimes called an "error of rejection" or a "false negative error." Denial of deserved bonuses? Demoralized employees? Lost talent for the organization? Law suits from employees?

The point here is that, to the extent that errors in classification can be tolerated, tests can be shorter. However, if the consequences of classification errors are severe, the tests used to make master/ nonmaster decisions will have to be longer, as well as meet other conditions required for reliable and valid tests. Chapter 11 explains how to calculate the reliability of a criterion-referenced test so that its adequacy can be examined.

Resources and Test Length

It will come as no surprise that the creation of tests takes time and, therefore, costs money. The longer and more sophisticated the test, the greater the development costs. There are also costs, of course, associated with maintaining and scoring tests. Test designers are perhaps less inclined to realize that tests also incur other costs to the organization—some dollar costs and others in the form of what are called "opportunity costs."

The dollar costs result from paying employees while they sit for tests. It is widely acknowledged that the greatest costs of training usually result from having to take employees off the job, while they are being paid, to participate in training. Testing results in similar costs. The opportunity cost of testing is time lost to instruction. If course designers have only two days in which to deliver a course, and test designers create a three-hour test, those three hours constitute precious time that cannot be used for instruction. It seems obvious that the more time, and hence money, that an organization can afford to spend on testing, the more reliable, i.e., the longer, tests they can afford. Organizations on a tight budget will need to trade off carefully the cost of test development and implementation against the cost of errors in test results. Once again, knowing the consequences of testing errors is essential to balancing this trade-off wisely.

Domain Size of Objectives and Test Length

The number of items required for a test is also influenced by the objectives that the test is designed to assess. In general, the smaller the domain of content described by the objective, the fewer the items required to assess the objective adequately. For example, consider an objective such as, "Without assistance, list the six steps required to make a milk shake using the Presto-Malt machine." This objective describes a small-content domain; in fact, it is difficult to imagine how one could write more than one item to assess this objective. Most objectives, however, require more than one item—parallel items—to assess them adequately.

For example, consider the following objective: "Given pertinent data and access to all essential technical manuals, diagnose the source of a radiation leak in a nuclear reactor." This objective describes a far larger content domain and, as you can imagine, would require far more items to instill confidence that it would be adequately assessed. Objectives that describe behaviors that must be performed under several different conditions on the job should be assessed by several items reflecting those different conditions. This discussion should make it clear why specific objectives are so important to test creation. It is very difficult to decide the issue of test length if the objectives are ambiguous.

Homogeneity of Objectives and Test Length

Another characteristic of the assessed objectives that influences test length is the homogeneity of the objectives, i.e., their relatedness to one another. Consider these two objectives: (1) "Without access to references, describe the steps in conducting a performance appraisal." and (2) "Without access to references, describe the four stages of interpersonal confrontation." These two objectives are related in that the content they cover is similar. In fact, the second objective is very likely a prerequisite objective to the first. As a result, test-takers are likely to perform the same way on the test items written for these two objectives; in more technical language, responses to items covering these two objectives will be positively correlated. If objectives are homogeneous to the extent that they result in test items to which test-takers respond similarly, fewer items need be included to assess each objective independently.

It is important to realize that it is very frequently difficult to tell simply by looking at objectives whether or not responses to their corresponding test items will be similar. This conclusion can only be drawn after actual test results are available and you can determine for certain how similar the responses were. Chapter 8 presents item analysis procedures that will explain how to make this determination. You may be able to reduce the numbers of items included for each objective on a test if you can confirm sufficient homogeneity of the underlying objectives. On the other hand, if the objectives covered by the test are largely unrelated—heterogeneous—you can expect that the test will have to be considerably longer since several items will probably be required for each objective.

Research on Test Length

Research into the accuracy of assessments as a function of numbers of test items per objective routinely indicates that more items result in greater accuracy. However, the improvement in accuracy tends to level off generally somewhere between four and six items per objective. This finding, however, refers to objectives in general without regard for the *criticality* of the objectives assessed. The accuracy achieved with four to six items may not be good enough for some critical objectives. In other words, the assessment of objectives that describe behaviors essential to health, legal requirements, and organizational survival should be assessed by more than six items, possibly by as many as 20 items, and may need to be assessed several times, especially if the content domain of these critical objectives is large.

Multiple assessments are more frequently used in performance testing than in paper-and-pencil testing; however, for some essential skills, multiple paper-and-pencil assessments may be appropriate. When using multiple assessments, the pattern of test-taker success and failure becomes important. See Chapter 10 for a discussion of the importance of consecutive success when multiple assessments are used.

Summary of Determinants of Test Length

The number of items that should be included on a test is primarily a function of the criticality of the master/nonmaster classifications that will be made based upon the test results. This is the case because test length is directly related to test reliability. The more costly the consequences of classification errors, the longer the test should be. Time and money, of course, are always limiting factors. Objectives that specify small content domains and that are correlated with other objectives require fewer items than those that describe large content domains and are essentially unrelated to other objectives covered by the test. Research suggests that the balance between effectiveness and efficiency in item numbers is achieved at four to six items per objective, but we know that more items will be required for some critical objectives.

You might use the numbers presented in Table 4.1 as a first estimate of how many items to include per objective on your test. Remember, however, that the actual reliability and validity of the test can only be determined after some test results have been collected. Chapter 11 describes the procedures for establishing the test's reliability and validity. In the absence of actual test data, you can only estimate how many items you will need to include on the test.

TABLE 4.1 Decision Table for Estimating the Number of Items Per Objective To Be Included on a Test

If the Objectives Are:			# of Items
Critical	From a Large Domain	Unrelated	10–20
		Related	10
	From a Small Domain	Unrelated	5–10
		Related	5
Not Critical	From a Large Domain	Unrelated	6
		Related	4
	From a Small Domain	Unrelated	2
		Related	1

5.

Performance Tests

FOUR TYPES OF RATING SCALES FOR USE IN PERFORMANCE TESTS (TWO OF WHICH YOU SHOULD NEVER USE)

Performance tests always involve the rating of a behavior or a product. A valid performance test is based on a detailed and thorough analysis of the skills required for the behavior, or the desired characteristics of the product, or both. Once the behavior (or final product) has been analyzed to define the essential characteristics of worthy performance, the next step is the creation of a rating scale to support a final evaluation. There are basically four types of rating scales: numerical, descriptive, behaviorally anchored, and checklists. Of these four, we do not recommend the use of numerical or descriptive scales as they allow for too much rater subjectivity. Both behaviorally anchored scales or checklists are acceptable approaches to assessing a skill or product, but of these two, the checklist is generally more reliable.

Numerical Scales

The numerical scale divides the evaluation criteria into a fixed number of points defined only by numbers, except at the extremes.

In other words, there is no definition of what level of performance merits a particular numerical rating, for example, what does a "3" mean on a 7-point scale? These pure numerical ratings are inevitably highly subjective assessments that can introduce substantial error into the testing process. Figure 5.1 illustrates an example of a numerical scale.

Descriptive Scales

Descriptive scales do not use numbers, but divide the assessment into a series of verbal phrases to indicate levels of performance. The descriptors may vary, for example, "Very Good" to "Very Poor," "Strong" to "Weak," but the resulting scale is deficient in that these words are open to many interpretations. Figure 5.2 provides an example of this second type of scale.

FIGURE 5.1 Numerical Scale

Behavior	Performance
	Poor Excellent 1 2 3 4 5 6 7
1. The quality of the response for directory assistance was.............	1 2 3 4 5 6 7
2. The statement of course objectives was...	1 2 3 4 5 6 7

FIGURE 5.2 Descriptive Scale

Behavior	Very Poor	Poor	Average	Good	Very Good
1. The quality of the response for directory assistance was.............					
2. The statement of course objectives was					

Behaviorally Anchored Numerical Scales

The behaviorally anchored numerical scale (sometimes called BARS) uses both words and numbers to define levels of performance. However, the words that are used are not vague value labels, but terms that describe specific behaviors or characteristics that indicate the quality of the performance or the product. The use of specific descriptions tends to make these scales more reliable than the unanchored numerical or descriptive scales. Figure 5.3 provides an example of this type of scale. As you can see, the more specific the behavior interpretation, the more reliable the scale will be.

One issue that often arises with the use of these scales is, "How many points should there be on a scale?" While the selection of points will be tied to the behaviors required for the task, research suggests that raters can reliably distinguish among five levels of performance. More than seven such points may stretch the limits of the rater's ability to quickly and accurately discriminate behaviors.

Checklists

Checklists are constructed by breaking a performance or the quality of a product into specifics, the presence or absence of which is then "checked" by the rater. Checklists tend to be the most reliable of all rating scales because they combine descriptions of specific

FIGURE 5.3 Behaviorally Anchored Numerical Scale

Behavior	Performance	Rating
1. Response to directory assistance request	1. Curt voice tone and inaccurate or delayed response	1
	2. Delayed response with neutral voice tone	2
	3. Average response time with neutral voice tone	3
	4. Shorter than average response time with pleasant voice	4
	5. Very short response time with very pleasant voice	5

FIGURE 5.4 Checklist

Behavior	Performance	
Statement of Objectives	Yes	No
1. The course objectives specify the action required of a student to demonstrate an ability to perform...............................		
2. The course objectives give the *conditions* under which the action will occur..............		
3. The course objectives specify the *criteria* by which the student and instructor will judge successful performance.........................		

behaviors or qualities with a simple yes-or-no evaluation from the rater. The checklist radically reduces the degree of subjective judgment required of the rater and thus reduces the error associated with observation. Remember, however, that while the checklist increases the reliability of the raters, a careful task analysis is required to assure the validity of the scale. Figure 5.4 is an example of a checklist.

PART THREE

PILOTING THE TEST

6.

Organization and Administration

PREPARING TO COLLECT TEST DATA

Before you can proceed with the final compilation of your test items, set the cut-off score that defines mastery, and establish the reliability and validity of your test, you need to administer the test on a trial basis. These remaining steps in the test creation process require test data from real test-takers. Therefore, you need to organize the test as similar as possible to its final form first.

BEFORE YOU ADMINISTER THE TEST

Giving the test is often viewed as a perfunctory part of the testing process. Administering the test often seems to be a matter of logistics that are inevitably complicated by the hassles of room location, duplicating, and scheduling. Unfortunately, it is all too easy to get bogged down in these "details" and forget that the test administration process needs to be viewed with the same philosophy that you have brought to bear on the entire test planning process, i.e., the goal of a systematic approach to test development is to reduce the error component (Xe) of test scores that arises not only from the test itself, but in the way the test is administered. Thus all those "details" need to be attended to and managed to reach our goal that the true score (Xt) equals our observed score (Xo).

Most of the issues we are now going to discuss apply to both paper-and-pencil tests and performance tests. However, there are some special considerations for performance tests, on which we will comment further into the chapter.

Sequencing Test Items

Imagine how you would feel if you were taking a test that started with the most difficult items first and, to make matters worse, switched back and forth between test item formats such as essay and multiple choice and fill-in. If you weren't demoralized when you began, you would soon be confused by the rapid succession of item formats. Demoralization and confusion result in testing error, and should be avoided. So, when you design the layout of your test, consider these two rules:

- *Keep item types together.* Don't mix fill-in items with multiple-choice items, for example. The ability to move back and forth between different types of item formats is the result of a particular personality type, not the test-taker's degree of content knowledge.

- *Arrange items starting with the easiest and proceeding to the most complex.* Typically, this means beginning with true/false and then moving to matching, fill-in, multiple-choice, short answer, and finally essay. This format prevents test-takers from getting bogged down on hard items at the beginning and running out of time to answer easier items at the end.

Three Types of Scoring Systems

With your test items arranged by difficulty and format, your next step would be to determine the type of scoring system you will use. There are two types of scoring systems that are traditionally used in paper-and-pencil tests: hand scoring and optical scanning (or OPSCAN). In addition, computer based testing systems, where the learner takes the test directly on a terminal, are now entering the market.

Hand Scoring. Most of us are familiar with the process of hand scoring a test. Hand scoring is easy for a short test given to a small number of test-takers. However, it can become quite tedious and fatiguing (and thus lead to error) when there are a large number of items and/or test-takers. To help reduce error in hand scoring, we suggest you consider using these techniques.

1. Save time in the scoring process by locating the space for the correct answer at the left-hand margin of the test, next to the item number. Your directions to the test-taker would be to "place the letter of the correct answer in the blank next to the item." This practice prevents your having to hunt all over the page for the test-taker's answers.

2. **Do not instruct** the test-taker to first circle the correct choice and then transfer the circled letter to the blank. Some test-takers will inevitably circle one letter and record another. Scoring controversies result when the test-takers assert that the circled answer was really what they meant, and that they just made a mistake in transferring the letter to the space.

3. When you develop the answer key, place the answer for each item so that it lines up with the corresponding blank on the left side of the test. Using a copy of the test itself to record the correct answers makes a useful key. All you need to do is align the blanks on the answer key with the test to score the test efficiently.

Optical Scanning. With the increased power of personal computers, OPSCAN systems are becoming more accessible. OPSCAN systems use a lens to read the placement of marks made by the test-taker on a special answer sheet and therefore rely on precise location of answers on the scoring sheets. An OPSCAN answer sheet must be printed in a very precise manner, along with the addition of special "timing" codes, and usually using a vegetable-based or low carbon (less than 40%) ink that won't interfere with the reading process.

While all of this may sound quite complex, OPSCAN scoring is a relatively inexpensive technology that is highly reliable. OPSCAN systems are fast (up to 10,000 forms an hour), accurate (virtually perfect), never tiring, and most of all, exceedingly informative. Once the computer has the test scores in memory, it can perform a number of analyses on the test that will help you interpret your test scores. In Chapter 8 we will look at some of the item statistics that you can easily create and use with an OPSCAN system. The Resources section at the end of the book lists vendors who provide OPSCAN systems and scoring packages. (NOTE: Many OPSCAN statistical packages are based on norm-referenced test assumptions. The use of these systems to interpret a criterion-referenced test can lead to a faulty test analysis. It is important to be certain that you understand which statistics within these packages apply to criterion-referenced tests [see Chapters 8 and 11]).

OPSCAN formats vary with different systems. As a result, it is very important that you make certain the test-takers understand how the answer sheet is to be filled out, and with what type of implement—usually a #2 or softer pencil that the machine can read.

Computer-Based Testing. While most training organizations primarily use hand-scored exams, and secondarily OPSCAN systems, a number of organizations have begun to use computers to deliver, analyze, and often provide feedback on tests. The computer-based testing system has a number of advantages. For example, Kelly Services, which provides temporary help, can assess keyboarding skills among job applicants for a number of different word processing systems while tracking error rate, speed, and accuracy by using a computer-based system to deliver their tests. Such data could also help diagnose areas where training for employees is needed. In addition, using such a system allows an organization to test an employee at any time and anywhere there is a proper terminal.

Computers also allow for an approach called Computerized Adaptive Testing (CAT), a process where the computer not only scores the exam as it is taken but also selects new items based on what is termed "item response theory." Item response theory is a statistical technique that selects items based on their difficulty, their ability to discriminate among test-takers of various competence levels, and the probability of guessing a correct response. CAT, which is a very new technique, is being developed primarily for traditional personality tests. As such there has been little use of the procedure in corporate settings. Computer-based testing in general is now entering the business arena, but as with computer-based training, it requires a relatively stable content base (since revisions tend to be expensive) and large numbers of users to justify its costs.

Problems in Correction for Guessing

As we pointed out earlier, the multiple-choice test is one of the most powerful techniques in test design. One of its weaknesses, though, is that test-takers do not have to create the right answer, they just have to recognize it. Many test designers wonder if there isn't some value in developing a correction for guessing to compensate for the fact that it is possible to get a certain percentage of multiple-choice items right simply by guessing. We believe that the disadvantages of correction for guessing outweigh the advantages, yet we know that test developers need to understand why, since they are frequently asked to apply such corrections. To understand the problems with adjusting scores for guessing, you need to know what the correction-for-guessing formula looks like and how to apply it.

If you give students a 100-item multiple-choice test where each item has four alternative responses, students will, on the average, get 25 items, or 25%, correct simply by guessing. (If there are five choices, they will get 20% correct, etc.) What the correction-for-guessing formula does is to subtract from the test-taker's number of correct answers the estimated number of answers that were answered correctly simply by guessing. The number to be subtracted is estimated from the number of items that the test-taker answered incorrectly. A correction-for-guessing formula thus assumes that every wrong answer resulted from guessing incorrectly. Figure 6.1 illustrates a correction-for-guessing formula and its application on a test with 95 items and five choices per item. As you can see from this example, the original score of 85/95 has now been reduced to

FIGURE 6.1 Correction-for-Guessing Formula

Score $= R - \dfrac{W}{n - 1}$

Where R = number right
W = number wrong
n = number of alternatives

Correction-for-guessing example

R = 85
W = 10
n = 5

Score $= 85 - \dfrac{10}{5 - 1}$

$= 85 - \dfrac{10}{4}$

$= 85 - 2.5$

$= 82.5$

82.5/95 due to the subtraction of a portion of the number of items attempted but answered incorrectly.

Although the correction-for-guessing formula seems to make good sense intuitively, research has revealed a problem with its use. As it turns out, very few responses to multiple-choice test items are pure guesses. Most of the time, the test-taker has some partial knowledge about each item and is making a decision based on more than random chance. In such cases, the correction-for-guessing formula penalizes test-takers for applying partial knowledge, since it is not guessing that has led them to select the wrong answer. Furthermore, people differ widely in their willingness to take risks—in this case, the risk is associated with making a choice about which they are not completely certain, and knowing they will be penalized if they are wrong. The result is that low risk takers refrain from answering many items that they would in fact get correct if they marked them. Therefore, the resulting scores of these persons are substantially lower than they should be. In other words, much current research indicates that the threat of the application of the correction-for-guessing formula introduces more error into the test scores than is eliminated by correcting for guessing. As we said, we do not recommend the adjustment of test scores by using a correction-for-guessing formula.

Test Directions

Your instructions to the test-takers must clearly describe, for each type of item, the action that is expected of the test-taker and any time or other constraints under which they must work. There are five common concerns in creating test directions:

- For multiple-choice questions, state very clearly that the test-taker is to select the "best alternative" rather than the "correct alternative." Failure to specify "the best alternative" will often lead to confusion on the part of test-takers. Good distractors are often "partly correct" but not "the best answer."
- List the time limits for the test, or for each section if appropriate.
- Describe the method of recording responses. If scratch work needs to be turned in, indicate where the work is to be done.
- Provide a cover sheet (or initial computer screen for computer delivered tests) that states the test's purpose. The cover sheet prevents test-takers from seeing the first test items while you

are distributing the test and from becoming distracted while you are discussing test procedures.

- Provide sample test items (and a few minutes in which to practice them) if the learners are unaccustomed to the test format.

Test Readability Levels

An often overlooked area in test development is the readability level of the test. If the readability level, either of an item or of the directions, is too high for the test-taker, you may be testing reading level rather than the competence specified in the course objectives. There are a number of methods to determine reading level, but they all focus on some relationship between words per sentence and the number of syllables per word. In general, reading level increases as sentence length increases. It also increases as you increase the number of multisyllabic words in each sentence. Conversely, it decreases as you decrease sentence length or the number of multisyllabic words.

The easiest way to examine the reading level of a test is to first analyze the directions for the test and then the items themselves. For multiple-choice items, analyze the stems and a randomly selected response if the response is in the form of a sentence completion. If the item stem is a question, analyze just the stem itself.

While there are a number of readability formulas, we have found that the formulas that calculate a reading level based on the average number of syllables per word are somewhat more realistic than those that use a "percentage of hard words" (words of three or more syllables). Therefore, indices such as the Kincaid or Flesch are probably better than the Gunning Fog Index. For example, the Fog Index for the knowledge level example on page 57 in Chapter 4 calculates to grade 24, while the Kincaid Index is grade level 11. The application level question on page 58 is about a grade 8 on the Fog and grade 5 using the Kincaid. If this were a two-item test, the average reading level would be grade 16 for the Fog and grade 8 for the Kincaid. As you can see, there are differences in the indices, and you do need to average a number of items before the reading level stabilizes.

Whatever you do, don't just run your exam through the reading index in your word processing program. The short phrases in item alternatives, fill-ins, etc., will lead to inaccurate estimates. Figure 6.2 provides a summary of the Kincaid Index procedures, which have been used extensively by the government. The index produces an approximate grade level based on the average syllables per word and

FIGURE 6.2 Kincaid Readability Index

Grade Level = .39 × (Average words per sentence) + 11.8 × (Average
  syllables per work) − 15.59

---

the average words per sentence. One final note: Reading level of an item has little to do with its difficulty level. As we just saw, a simple knowledge level item can rate much higher in reading level than a more complex application item.

Formatting the Test

Finally, as you organize and prepare the test for administration, you should review it for editorial details that can affect error in testing. Ask yourself these questions:

- Are the directions clearly identified before each section?
- Are the items spaced to avoid crowding?
- Are questions and responses physically linked so that the test-taker does not have to flip back and forth between pages (or computer screens) to review the question being answered?
- Are the print or electronic characters legible? (Never use less than a 10-point font.)
- Are diagrams neat and accurate?
- Has the test been proofread? (One technique is to read the test from back to front—it's painstaking but generally accurate.)

Setting Time Limits—Power Versus Speed

Trying to decide how much time to allow for a test can be tricky. However, when doing so, you should first determine whether the test is to be a power test or a speeded test.

Power Tests.   Most of the tests we have taken in our lives have been power tests. A test to assess mathematical ability or one to determine knowledge of supervisory skills is usually a power test. These tests are designed to assess what you know rather than how fast you can demonstrate what you know. Power tests usually have liberal time limits that assure that most people will finish the test. In theory, a power test would have no time limit, but in practice we usually need to put a limit on the allotted time. It's hard to estimate, but you might begin by allowing at least 45 seconds for each question, and possibly two minutes for more difficult items.

Speeded Tests. When you have a situation where you must consider how quickly a skill is performed, for example, a series of physical actions that must be completed in harmony with an assembly line, or a clinical judgment about a severe drug reaction, then you need a speeded test. These tests are carefully timed, and the inability to perform satisfactorily within the time limits should result in a nonmastery decision. Time limits for a speeded test should be set to match the time limits for performance of the assessed skill on the job.

In the end, the length of time for the test should be a data-based decision. You will need to pilot the test, and during the pilot, you should make certain that your time limits are appropriate. Thus, for power tests, you must budget enough time to assure that most test-takers can complete the test. If you don't, you convert it *de facto* into a speeded test with a probable increase in erroneous nonmastery decisions—not to mention an increased probability of lawsuits.

WHEN YOU ADMINISTER THE TEST

Test administrators have a responsibility to make sure that each test-taker has an equal opportunity to demonstrate what he or she knows. To help the test-takers to perform at their true levels, it is important to control both the physical *and* the psychological environment.

Physical Factors. When we first introduced the idea of error in testing, we listed a number of conditions that could reduce a person's true score. Factors such as room temperature, humidity, ventilation, noise, workspace, and lighting should all be checked— and adjusted if need be—right before the test is administered, to make sure they help contribute to test performance, not detract from it.

Psychological Factors. Most testing situations will create a certain level of anxiety for the test-takers. We know from research studies that people perform best when their anxiety level is neither too high nor too low. As a test administrator, you will need to establish an affective environment where test-takers see the test as serious, but not excessively threatening. The best advice for controlling the psychological factors associated with the test is to act in a professional manner. You should be calm, friendly, but work oriented. There has been some discussion in the literature that performance is facilitated with a same-sex or same-race test administrator. If you wish to be especially sensitive to the psychological environment of the test, you might want to consider this. Matching administrator demographics to the test-takers may be somewhat extreme in most instances, but if your test will be affecting employment decisions regarding a group of people who are primarily in a protected minority, it might be good test policy to select a test administrator with similar characteristics to administer the test to them.

Giving and Monitoring the Test

Giving proper test directions is an integral part of the testing process. The test designer should clearly specify not only what the test-taker is to do during the test, but what the test administrator needs to do as well. In the course of the test administration there are three stages you should attend to: becoming familiar with test directions, giving test directions, and monitoring the test.

1. *Familiarize yourself with the test and its required procedures before the testing begins.* If you have never administered the test, or if it has been awhile since your last administration, then you should make certain that you understand what is required of the administrator. As test administrator you should obtain a checklist to make sure that test-takers have the appropriate support materials they need for the test, *but nothing more.* All should have equal access to reference materials if they are allowed or required for the test.

2. *Refrain from unnecessary talking before the test.* Most test-takers will be anxious, if not eager, to get on with the test. They are not likely to be in a frame of mind to listen to irrelevant chatter or information about future coursework. Right before a test is perhaps the worst time to deliver any information, not pertaining to the test, that you want the test-takers to remember.

(Under no circumstances should you administer a course evaluation before a test!)

3. *State directions* exactly *as prescribed by the test designer.* Giving test directions differently to different groups of test-takers can result in test scores that are not comparable across groups. Doing so can result in not only imprecise testing, but also litigation.

4. *Make sure that all test-takers know how to take the test.* Right after you have given the directions, ask if there are any questions, and then pause long enough after asking to allow people who are thinking about what you said the time they need to respond with their questions.

5. *Keep talking to an absolute minimum.* As administrator you need to be available to deal with procedural questions, but these conversations should be as short and as quiet as possible. If a test-taker discovers a flaw in an item that might cause confusion during testing, the confusing point must be clarified for the entire group of test-takers, as soon as possible after its point of discovery. Providing clarification only to the person who discovers the flaw will give that person an advantage over the others. Also, talking to individual test-takers as the flaw is noticed will lead to a constant sense of distraction during the test.

6. *If someone asks about a question, do not provide hints to the answer.* Unless the item is genuinely flawed and in need of clarification for the whole group, instruct those who question to use their own best judgment. Remember that dropping hints is unfair, introduces error, and can lead to grievances or litigation.

7. *Do not allow talking among test-takers during the test (unless the test permits such interaction).* While it's common sense not to allow talking during the test, many administrators will forget that they too shouldn't be talking during the test. It is easy to become bored with the monitoring function and to end up whispering to a colleague in the back of the room. However, such behavior can easily become distracting to anxious test-takers. If you expect to have time on your hands as a test administrator, we suggest taking some light reading to occupy your mind. However, test administrators should remember that their role is an active one; they must remain alert to any factor during the test that would introduce error in the scores.

8. *The test should be monitored from different points in the room.* Moving around will help to ensure that the directions have been understood and are being followed. Attention to nonverbal cues

can help identify confusing points even before a hand is raised to seek clarification; however, don't stand over someone's shoulder. Many people find this behavior intimidating and anxiety producing.

9. *Adhere to time limits.* Consistency is fundamental in testing. If time limits have been specified, follow them. If you discover that the test-takers are having trouble completing the test as you expected, make a note for the test designer, don't modify the situation on the spot. There are too many interrelated variables in a testing situation to expect that a modification in one part of the system won't affect other parts.

Special Considerations for Performance Tests

Most of what we have just discussed applies to performance tests as well as paper-and-pencil tests. Still, there are some special issues associated with conducting the performance test. These issues affect before-test planning and test monitoring.

1. *Carefully review the rating form.* The last thing you need during a performance test is to have to ask the test-takers to slow down so you can figure out how to use the rating form, or to ask them to repeat a performance because "I didn't see this other section." The rating form review will be especially important in those settings where a series of rapid decisions and actions are likely. (As an instructional strategy, we also believe that students should have received a copy of the rating form in advance of the test. This will not only allay anxiety, but help the learner to focus on the competencies assessed in the test.)

2. *Determine that the test environment simulates the work environment.* With the rating form in hand, review the test setting to be certain that whatever resources are to be available for the test, for example, electronic test equipment, reference manuals, etc., are in place. However, don't just review for the test itself. If you find there is a significant disparity between the work environment and the test environment, you should note this difference in the event that such a gap is a concern for the test-takers. If the gap is too large, then issues about the test's validity may be raised, for example, a simulation of air traffic direction without constantly moving radar images would be a serious breach of validity for air traffic controller assessment.

3. *Plan the schedule for testing.* Once you are comfortable with the rating form and the environment, determine how long it will take to test each person. The test administrator needs to develop a testing schedule that will minimize the effects of anxiety, and possibly the unequal levels of instruction that result when some test-takers are allowed more study or practice time than others.

4. *Test-takers don't appreciate irrelevant assignments immediately before a test.* If you are in a setting where there is limited equipment and/or raters to conduct the test, you should not keep the test-takers sitting around in the classroom area while they wait for the exams; this will only create anxiety or boredom. Nor should you provide additional instruction or opportunities for practice once testing has begun; this would be unfair to those who were tested first. Rather, the best approach would be to assign each individual a specific time to be tested. You then allow the test-takers to use the balance of the time as they wish—to study, jog, call the office—both prior to taking the test and while they wait for others to be tested after them.

5. *Monitor the setting.* During testing you may have a situation where you can't schedule people one at a time. If you have to test several people at the same time, in the same space, be sure to spread them out as much as possible. Failure to do so will increase the likelihood of unwanted interaction among test-takers. If you don't want to make explicit your concern about unfair assistance, you may emphasize that the distribution of testing areas is to provide test-takers with enough room to work and to avoid distractions (which is also true).

HONESTY AND INTEGRITY IN TESTING

Honesty and integrity in testing have more than an ethical significance. Their lack contributes to the error component of test scores. We all realize that testing can be a stressful event, whose results may well affect employment or promotion. As a test administrator you may find yourself in the uncomfortable position of resisting pressures "to help" a friend, colleague, or even worse, an important executive. Clearly, a testing system needs to be insulated from these pressures for ethical, political, and legal reasons. You should take precautions to protect the integrity of the test not only during a training-testing sequence, but on an organizationwide basis.

Security During the Training-Testing Sequence

On a day-to-day basis when testing follows training, there are three safeguards you should use against these sources of error.

Test Item Security. Allowing some test-takers to see the items in advance is obviously unfair to the others. An instructor might feel that it's "OK" to talk about the test to a whole group, since that "wouldn't be unfair to any individuals," but this practice is still unadvisable, for two reasons.

1. Such an action will be unfair to other groups of test-takers who were not allowed to preview the test.

2. A preview may well destroy the validity of the test items. While it is sometimes felt in objectives-based testing circles that if instruction is based on a mastery model, then test-takers ought to be able to see the test in advance. However, whether or not this is the case depends entirely on what kinds of objectives are assessed by the test. For instance, suppose you are testing a particular skill that requires test-takers to identify unassisted a previously unseen example of a concept. For those test-takers who see the example in advance, the item probably will be reduced to mere recall during the test administration.

If we were testing to see if TV technicians could, without help, correctly label previously unseen examples of different kinds of video shots, the validity of these test items would be destroyed for any test-taker who was allowed to see the examples in advance of the test, even if answers were not provided. Seeing the examples before the test might allow the test-taker to seek assistance with the answers. Providing answers to such items means that test-takers have to *remember* answers during the test, which is not the same as classifying examples without help.

Interaction among Test-takers Should Not be Allowed. Unless the test requires group interaction, for example, a cockpit flight simulation, then the test-takers should not be allowed to talk during the test. Nor should they talk about the test afterwards. During the test it is often difficult to enforce this "no talking" rule with adult learners. The best means to do so, as we discussed earlier, is through professional and serious demeanor. If your behavior and attitude demonstrate to the test-takers that you consider testing a serious process, you are less likely to have problems in winning the cooperation of the group.

After the test, the test-takers should be encouraged not to share information about the test items. Such sharing means introducing error into the testing process, i.e., comprehension items reduced to recall as described above. The test-takers may be under pressure from colleagues to divulge information about the test. In this instance, a caution to the test-takers about the ethical or measurement issues surrounding sharing information may not be as powerful as a reminder that others are likely to outscore them if they know more about the test items.

For Placement Purposes, Test-takers Must Do Their Own Work. In many instances a test may be offered at the test-taker's work site rather than in an instructional or formal test setting. Usually these tests involve an assessment for course equivalency, entry, or prerequisite skills. When these tests are offered at a work site, there is a real opportunity for supervisors or others to provide assistance to the test-taker. It is most important that you communicate with the test-taker and his or her supervisor to emphasize the importance of achieving valid test results for placement purposes.

Organizationwide Policies Regarding Test Security

Most organizations that are concerned about test policies concentrate on three points: security, access, and destruction.

Security of the Test. Test materials should always be inventoried and materials kept in a physically secure area. Obviously, tests should not be left out where they could be seen. Any requests for tests that are to be administered outside of the immediate training or testing areas, for example, an equivalency test mailed to a regional training office, should be logged and transmitted in a secure manner. An overnight delivery or a telefax can be received by any number of people. Be certain who is on the receiving end!

Access to the Test. A clear policy should be established and adhered to regarding who will have access to a test. A log system should be established that will provide for documentation of access to the test. If the test is available via computer, standard security measures, for example, passwords, should also be implemented to limit access. Tests shouldn't be made available to anyone not on the authorization list. Requests for access by those who are not authorized should be handled via prearranged procedures. Especially sensitive requests, for example, unions or regulatory agencies, should be referred to the legal department before any action is taken.

Destruction of Tests.   An organization should have a policy on retention of tests, for example, how long an individual's test should be kept in the event you need to provide evidence of performance in a legal challenge or a grievance. However, tests may need to be destroyed in the event:

- test forms or answer sheets have been written on,
- test forms are worn-out,
- test copies are defective or incomplete, or
- tests are outdated as courses are modified, superseded, or replaced.

In these instances be certain that the tests you sought to destroy are truly irretrievable. As any teacher will tell you, tossing an exam copy in the trash is no guarantee that the test has been taken out of circulation. Tests should be shredded or otherwise destroyed in the presence of a witness and so documented.

7.

Collecting Pilot Test Data

WHY PILOT A TEST?

Just as any systematic approach to course design includes a formative evaluation or course pilot, a systematic approach to test development means piloting your test. When designing a test, piloting is absolutely essential because the detection of faulty items requires real test data. The piloting process should identify potential problems with test organization, directions, logistics, and scoring as well as with individual items, and lead to their correction. Additional test data gathering will also be required in order to establish the cut-off score that defines mastery (see Chapter 9) and to establish the reliability and validity of the test (see Chapters 10 and 11). *The single most important purpose in the initial piloting of the test is to gather feedback for improvement of the test, not to rate the pilot test-takers.* Remember, almost any testing situation can be personally threatening. As you conduct the test pilot, you need to be particularly supportive and emphasize that your purpose is the evaluation of the test, not the test-takers.

SIX STEPS IN THE TEST PILOT PROCESS

The pilot test is a formative evaluation process that will parallel the course pilot process. You will need to determine the sample of test-takers, orient the participants, give the test, analyze the test results, interview the test-takers, and synthesize the results. You may not be able to do all of these steps as the result of a single pilot test administration, but by adhering to these guidelines, you should be able to assess accurately the quality of your test.

Determine the Sample

Your pilot test-takers should mirror your intended test audience. Don't rely on a "sample of convenience" where you grab three people who are around the office and in between projects. Nor should you be satisfied with just anyone sent to you from the field. If the pilot is to have meaning, the sample test-takers must be representative of future test-takers.

The size of the pilot test sample will depend on the scope of the test. A small sample will be useful primarily to gather qualitative reactions from the test-takers, i.e., verbal comments about how the test might be improved. If you are designing a test for a limited-run workshop, then you could work with a smaller sample—even as small as three. If you are designing a test that will be used on an ongoing, companywide basis, then you should invest the time and resources for a full pilot. So the logical question you should ask yourself is: How do I decide who I should pick for my pilot?

- If you have a test that will be used on a limited basis and you aren't concerned about legal or grievance issues, then you could go with a sample as small as three people (five would be better). We usually prefer that the three people be chosen to represent above-average, average, and below-average performers. This way we mirror, in a loose way, the total possible sample. If you have a choice of more than one person in each skill range, pick someone who is analytical and verbal over someone who would rather not talk—you will get more and better feedback.

- If, on the other hand, you are trying to select your sample for a test that will have greater consequences, then you should choose a representative sample that will at least mirror the normal class size that might be associated with the test. Generally speaking, this means a test pilot of 12–15 people.

- Don't try to establish test reliability or validity during an initial pilot of the test. The test is not likely to be ready for such measures at the pilot stage. The pilot process should always be completed in advance of these validation procedures (see Chapters 10 and 11). If you are under time pressure to complete the reliability and validity work, at least pilot the test on a small sample before bringing larger numbers together for the validation measures. You can imagine the frustration and wasted resources if you have selected a large test validation sample, planned to pilot the test and set the cut-off score at the same time, and then discover that words are missing from questions, questions are missing from the test, or the raters don't know how to use your checklist.

- Finally, be sure to document the characteristics of the pilot test group. The sampling decisions you make for the pilot should be noted in a memo or other written form in case questions arise later about the test.

Orient the Participants

Since your goal is to evaluate the test, not the test-takers, you need to make it clear to your pilot sample that it is the test that is being evaluated. Since the test-takers are, in effect, your colleagues in the test development process, they should be so informed and treated as such. You should begin the pilot process by establishing rapport with the test-takers and setting the collegial tone that will be needed to complete the pilot. Finally, be specific in terms of what you want the test-takers to do and what you will be doing during the test, for example, "I would like you to take the test just as it is, but make a note any place on the form where you were confused or had a question. I'll be circulating during the test to see if there are any problems, but I don't want to get involved in interpreting test directions or items unless I have to."

Give the Test

When you give the pilot test, give it exactly as it would be given in the field. This means you should give the directions verbatim, adhere to the time limits, and avoid any hints, apologies, or interpretations of the test or any of its items. Any intervention on your part during the pilot may jeopardize your understanding of how the test will work later in the field. Smaller errors such as typos in the test should be corrected just as you would during a field administration. Gross errors may require immediate modifications to the test in order to allow the pilot test-takers to proceed. In either instance, it is important that you document your changes and the reasons for the changes.

While the pilot test is in progress, you should take careful notes to document what the test-takers are doing. Watch for nonverbal cues such as head scratching or frowning, which might indicate anger, confusion, etc. Don't allow any interaction between test-takers unless it is called for in the test design. In general, do the same things we discussed in the section on monitoring the test (Chapter 6), but with added emphasis on watching for any problems with the test. Don't forget that you can exert a fair degree of control over the group simply by maintaining a professional demeanor.

When the test is done, be sure to thank the participants. You will still have other interactions with them during the pilot process, but they have probably worked hard during the test and deserve your recognition.

Analyze the Test

In an ideal setting, right after you administer the pilot test, you would complete the statistical analysis of the results. The test item analysis process allows you to identify any items that might be a problem, e.g., nobody selected three of your distractors on an item, effectively converting it from a multiple-choice item to a binary-choice item (with a fifty-fifty chance of getting it right by guessing). The data you gather from this analysis will then help guide you in the next stage of the pilot process—your interviews with the test-takers.

Chapter 8 provides a detailed discussion of the statistical process of analyzing a test. These techniques are most commonly and easily applied to multiple-choice tests—though they can be used with other types of tests. We will summarize the major techniques now:

- Difficulty index. This is simply a report of the percent of test-takers who answered an item correctly.

- Distractor pattern. This statistic is a report of the number of test-takers who selected each alternative option for each test item.

- Point-biserial correlation. A more sophisticated technique, the point-biserial correlation really requires computer support. It is, however, a very powerful tool that easily allows you to identify items that test-takers with the highest scores consistently missed while low-scoring test-takers consistently got right. Such items are generally poorly written and require modification.

These techniques take very little time to complete with computer support, and the first two can be done by hand with a little advance planning. The results can then be reviewed and serve to guide your interviews with the test-takers, for example, "How come you didn't select any of these three options? Were they too far off? What would be a better choice for a distractor?"

Interview the Test-Takers

After the test you should interview the test-takers. We recommend that you conduct interviews on an individual basis. You should plan your interviews based on two sources of data: your observations during the test and the test analysis data. When you begin each interview, first remind the individual test takers of the formative nature of this experience and thank them for their cooperation. Then continue with your questions. Referring to the testing session, you might ask about difficulties they may have had, for example, "I saw you scratching your head at question nine. Was that one a problem? Was it wording or content?" If your analysis has identified problems with specific items, ask test-takers how they interpreted the item. As they talk, take careful notes. Don't concentrate only on their performance either; be sure to explore their feelings about the test, for example, "How do you feel about this test? Would it be a fair test?"

A formative evaluation of the pilot test should be pursued in the same manner as any other formative evaluation. Don't use a series of closed-ended questions, but probe the responses and summarize test-takers' comments, to make sure you understand their thoughts. Whether they have volunteered or been volunteered for this experience, you should be sensitive to the fact that they may be anxious about the real purpose of the test. If you have developed rapport with the group and established that you are assessing the test and not the individual group members, your note taking and discussion should proceed without difficulty.

Synthesize the Results

While your impressions are still fresh, you should synthesize your findings and document them. If your organization uses a standard form for course pilots, you might adapt it to meet your needs for testing. Some of the standard information to include would be the following:

- Time of the test
- Location of the test
- Administrator
- Description of the participants
- The range of times it took to complete the test and the average time
- Any instructions or procedures that need to be modified
- Any test items that need to be modified
- Any format changes
- Any materials that need to be added or are unnecessary
- Overall impressions
- The item analysis report

PART FOUR

EVALUATING THE TEST

8.

Interpreting the Test Results and Item Analysis

STANDARD DEVIATION AND TEST DISTRIBUTIONS
THREE COMMON ITEM STATISTICS IN ITEM ANALYSIS

STANDARD DEVIATION AND TEST DISTRIBUTIONS

In this chapter we want to discuss how you can interpret the results of a test and use individual item statistics to improve the quality of the test. Before we describe and illustrate these item analysis techniques, however, there is one more concept that anyone trying to interpret a test must understand: the standard deviation. The standard deviation will tell you how spread out your test scores are (did everyone tend to get the same score or is there a diversity of scores?) and can affect your interpretation of test data.

The Meaning of Standard Deviation

After a test has been given, you will want to examine the scores. Individual scores sometimes have great significance to individual test-takers because a score can determine a career path or other reward. However, the analysis of the scores as a group will likely be of greater interest to course designers and test creators. The results of this analysis will tell you a great deal about the quality of your test and provide guidance for test revision before you attempt to establish the test's reliability and validity.

The first step you want to take in examining your test results is to construct a frequency distribution of the test scores. The reason for plotting your scores is that the shape of the distribution can tell you something about how test-takers did on the test. If, when you plot your scores, the distribution that results looks like Figure 8.1, then the test results are said to be normally distributed.

This bell-shaped curve is defined by the average deviation of the scores from the *mean;* the mean is simply the average score. In Figure 8.2 you can see that the curve is divided into sections. Each of these sections represents one or more standard deviations from the mean. In this instance, the average score or mean score is 50 and the standard deviation is 15 points. The "0" point on the standard deviations line represents the mean score. One positive standard deviation from the mean is a score of 65, and one negative standard deviation from the mean is a score of 35. In any standard normal curve, there will always be approximately 34% of the people scoring in each of the first standard deviations, about 14% in the second, and around 2% in the third.

FIGURE 8.1 Standard Normal Curve

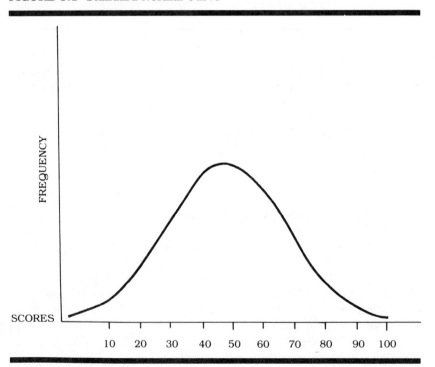

FIGURE 8.2 Standard Normal Curve with Standard Deviations

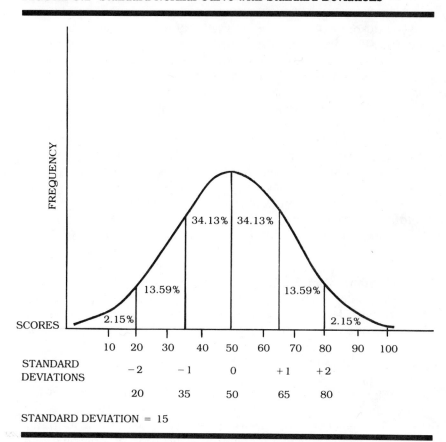

STANDARD DEVIATION = 15

Technically, the standard deviation is the average deviation of the scores from the mean or average score. Calculating a standard deviation is not complex, but it can be time consuming. Any worthwhile test scoring program will include this statistic automatically. From our perspective, it's not terribly important that you know how to calculate the standard deviation. What is important is that you understand conceptually that the standard deviation is a measure of how widely the scores are distributed about the mean, and that the percentages of scores (about 34%, 14%, and 2%) that fall within the standard deviations from the mean define the normal curve.

The Five Most Common Test Distributions

Any normal curve will be symmetrical in shape, and as we just saw, the standard normal curve has specific percentages of scores within a given standard deviation. Now, let's look at the impact of standard deviations upon the shapes of curves. In Figure 8.3 there is a sequence of three curves. The first curve is like the one we have just seen—a standard normal distribution that typically results from the administration of a norm-referenced test. This is also called a *mesokurtic* distribution. The next curve is one with a larger standard deviation. Notice how the scores are more spread out and the highest point on the curve is lower than in our first curve. (This curve is still a normal curve as long as the correct percentages of scores fall between the various standard deviations.) This flatter curve is called a *platykurtic* distribution. The final curve, a *leptokurtic* distribution, is one with a smaller standard deviation. Notice how the scores are closer together. The smaller standard deviation indicates less variation in the scores. Consequently, the highest point of the curve is much higher than in the other distributions.

You should be able to see from these examples that the smaller the standard deviation you obtain, the narrower the curve will be. Thus, a leptokurtic distribution will have a smaller standard deviation, indicating that most people tended to score alike. Mastery distributions will tend to be leptokurtic, but they also have another characteristic—skew.

In Figure 8.4 the top curve is a typical mastery curve. In this curve, most test-takers' scores are clustered near the high end of the scale. This clustering toward the extremes of the distribution is called the *skew*. Skewed distributions are not symmetrical and, therefore, are not normal distributions. The mastery curve is called a negatively skewed curve because the "tail" of the distribution is toward the low end of the scores. The bottom curve is a positively skewed curve. Notice that the skew is labeled positive or negative based on the direction of the "tail." Thus, a "tail" to the right is a positive skew, a "tail" to the left a negative one. A positively skewed test distribution is usually cause for test developer depression, as it represents a situation where most test-takers have scored poorly on the test.

Problems with Standard Deviations and Mastery Distributions

Some test designers consider a normal distribution of test results an indicator of a good test. These test creators have confused

FIGURE 8.3 Frequency Distributions with Standard Deviations of Various Sizes

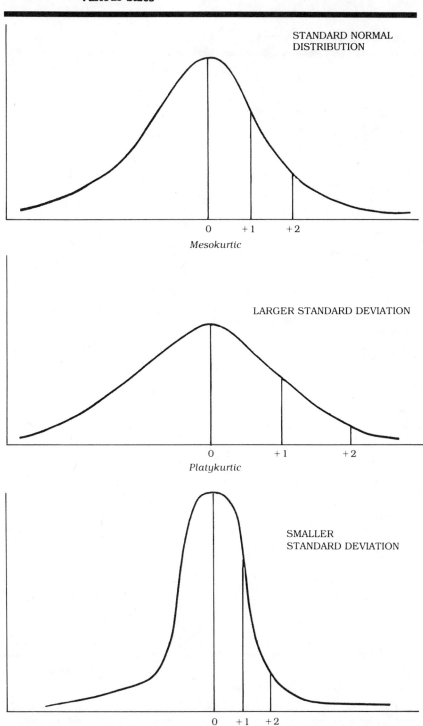

STANDARD NORMAL DISTRIBUTION

0   +1   +2

Mesokurtic

LARGER STANDARD DEVIATION

0   +1   +2

Platykurtic

SMALLER STANDARD DEVIATION

0   +1   +2

Leptokurtic

FIGURE 8.4 Skewed Curves

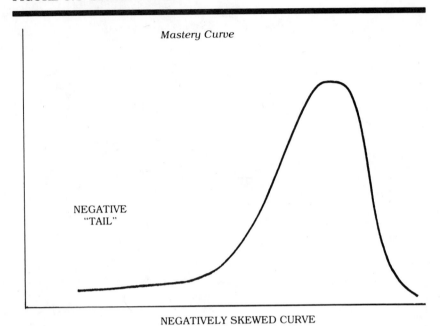

Mastery Curve

NEGATIVE
"TAIL"

NEGATIVELY SKEWED CURVE

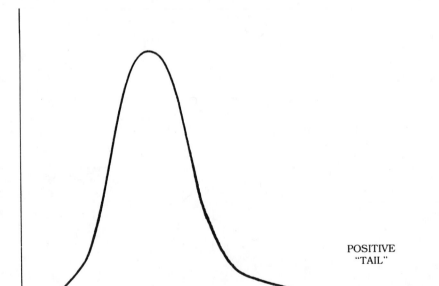

POSITIVE
"TAIL"

POSITIVELY SKEWED CURVE

norm-referenced and criterion-referenced test philosophies. When creating an NRT, the normal distribution is highly desirable because it indicates spread in test-takers' scores—the spread that is essential to the reliable ranking of test-takers against one another. However, if your goal has been to create instruction based on specifc competencies, and learners have largely mastered your objectives, your test distributions should look similar to Figure 8.5. In this test, the mean is represented by the "0" on the horizontal axis where the standard deviations are listed. In a skewed distribution like this, you can calculate a standard deviation, but the result won't be meaningful because the percentage of scores within each standard deviation won't be like those of a normal curve, and in addition, will not be the same on both sides of the mean. Remember, most of the traditional test statistics (point-biserial correlations presented in this chapter and internal consistency measures of reliability discussed in Chapter 11, for example) were designed for use with normal distributions. You can use some of them to guide you in developing a criterion-referenced test, but they are not as meaningful with nonsymmetrical (skewed) distributions. In most instances in-

FIGURE 8.5 Mastery Curve

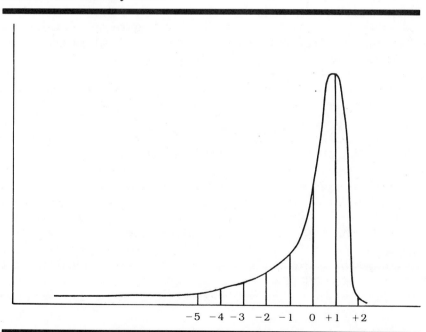

volving CRTs, you can get a good feel for the standard deviation by plotting the scores. If most of your students have mastered most of your objectives, you should expect a negatively skewed distribution with a leptokurtic (tight) grouping of scores—this would be a distribution with a small standard deviation.

THREE COMMON ITEM STATISTICS IN ITEM ANALYSIS

Once you have developed and administered your test, you need to review the test item statistics to help decide whether or not the test is "good." The process of reviewing each item and the way in which it contributes to the value of the test is called "item analysis."

Difficulty Index

The difficulty index is easy to understand once you realize it really should have been named the "easy index." The difficulty index is simply a measure of the number of people who answered a given item correctly. This statistic is usually expressed as a decimal, for example, .80 means 80% of the people taking the test answered the item correctly. The difficulty index can range from .0 (nobody got the item right) to 1.00 (everyone got the item right).

If you have a true mastery situation, you would expect to find high values for the difficulty indices since you expect everyone to pass. However, an alternative explanation for a high difficulty index is that the item is too easy or that people succeeded because of a technical flaw, for example, unrealistic distractors in a multiple-choice item. Because of these problems you should never use the difficulty index alone to decide about the quality of an item, as the value of the index varies depending on the test philosophy. In a norm-referenced test where you want to separate the test-takers, you would include items that ranged in difficulty from .30 to .70. Difficulty is not usually a major factor in deciding whether or not to include an item on a CRT. The decision is based on whether or not the item accurately measures achievement of an objective covered by the test. However, if a CRT is used as a part of instruction driven by a true mastery philosophy, you would expect most items to have an index of .90 or higher.

Distractor Pattern

When examining responses to closed-ended items, you should also look at the test-takers' responses to your distractors. If you find that test-takers consistently eliminate a choice as wrong, then your distractor may be too easy. If this happens, then you may have changed the odds of guessing the correct answer. Where you might have expected test-takers responding to multiple-choice items to have a one chance in four of guessing the correct answer, two generally unused distractors will shift those odds to one out of two—a shift from a 25% to a 50% probability of guessing the answer. In effect, your multiple-choice question has become a binary-choice assessment.

Point-Biserial Correlation

The point-biserial correlation is probably the single most useful statistic in the item analysis process. (If you are unfamiliar with the concept of correlation, you should read the section *Correlation* in Chapter 11 before trying to interpret point-biserial coefficients.) The point-biserial coefficient correlates the test-takers' performance on a single test item with their total test scores. What this means is:

- Each item will have a single point-biserial coefficient—a number—ranging from $+1.00$ to -1.00.

- A positive coefficient means that test-takers who got the item right generally did well on the test as a whole, while those who did poorly on the test as a whole generally missed the item.

- A negative coefficient is an indication that those test-takers who generally did well on the test missed the item while those who generally did poorly got the item right. (Items with negative point-biserial coefficients are not useful for developing sound norm-referenced or criterion-referenced tests.)

- If all test-takers answer an item correctly or incorrectly, that item will have a point-biserial coefficient of 0.00.

If the point-biserial seems to be confusing, it isn't. It is just a particular correlation technique that follows the patterns of any correlation calculation, as described in Chapter 11. The point-biserial is a coefficient designed to measure the relationship between two variables, one of which is dichotomous (has only two values) and the other of which has a continuous range of values. In this application of the point-biserial correlation, the dichotomous variable is performance on a given item; the two values are "correct" and "incorrect." The variable with the continuous range of values is the total test score.

As a result, any item with a negative point-biserial correlation is immediately suspect! If the poorest performers got an item right that the best performers missed, something is usually wrong. The most probable sources of the problem are:

- Some type of systematic misinformation is being disseminated in the instruction or the workplace.

- There is something about the item that causes those who know more to miss it, usually a subtle interpretation that the test designer missed in creating the item, but which better performers read into it.

- The sample size on which the item analysis is based is too small. This is not to say that under this circumstance the point-biserial incorrectly represents the relationship between test item performance and total test performance for this sample. The relationship is portrayed exactly as it is; however, if the sample size is smaller than 15–20, the negative finding might be due to chance error—some test-takers incorrectly coding a response or not paying attention—and might disappear as more test-taker data is collected.

When you review your tests, you will probably find that most items don't approach either extreme in point-biserial values— +1.00 or −1.00. This is especially so in mastery testing where one expects most performers to do well. The lack of variation in test scores will limit how high the correlation figures will be. *Regardless of the size of your point-biserial coefficient, review every item with a negative or zero correlation for possible revision.* Keep in mind, though, that if every test-taker gets an item correct or incorrect, the item will have a point-biserial coefficient of zero. Thus in cases where the point-biserial is zero, you will also need to look at the difficulty index to determine how test-takers did and whether or not their performance represents trouble. For example, most test designers (course designers, too) would be more troubled if everyone missed an item than if everyone got the item correct.

Garbage In/Garbage Out

Finally, remember that an item analysis, like any numerical technique, is only as good as what goes into it. As they say, Garbage In/Garbage Out. The item analysis won't tell you anything about the quality of the objective being measured by the item. It doesn't tell you how accurately an item assesses a given objective. You could have an item with a high point-biserial, full use of the distractors, and a high difficulty index on a test of supervisory skills when the item is really measuring reading comprehension. Once again, there is no substitute for competent professional judgment in the testing process. An item analysis package will analyze response patterns for the most trivial as well as the most crucial items.

PRACTICE

Review the following sample test item analysis. What are the three worst items?

Item	Difficulty Index	Point- Biserial	Distractor Choices				
			1	2	3	4	5
1	1.00	.00	6*	0	0	0	0
2	.17	.75	5	1*	0	0	0
3	.33	.55	2*	0	1	3	0
4	.83	-.88	5*	1	0	0	0
5	.67	-.19	2*	4	0	0	0
6	.33	.55	1	1	2	2*	0
7	.50	-.71	3	3*	0	0	0
8	1.00	.00	6*	0	0	0	0
9	.67	.46	4*	0	2	0	0
10	.67	.22	4*	2	0	0	0

*Indicates the correct response

Number of Test-Takers = 6   Low Score = 2
Standard Deviation = 1.79   High Score = 10
Standard Error = .41 (see Chap. 9.)   Average or mean = 5.8
Internal Consistency Reliability Estimate = .16 (see Chap. 11.)

FEEDBACK

The three worst items appear to be 4, 5, and 7. All have negative point-biserials, and only two distractors have been chosen in each of these items.

9.

Standard Setting

DETERMINING THE STANDARD FOR MASTERY

One of the most difficult, yet critical, tasks required in CRT development is to determine the standard for passing, i.e., the cutoff score that separates masters from nonmasters. The testing literature presents several methods for doing this, and we will describe three of them. Before we look at these procedures we want to make you aware of a number of considerations that affect the standard-setting process, regardless of the method used.

THE OUTCOMES OF A CRITERION-REFERENCED TEST

Following the assumptions of criterion-referenced testing, the true status of every test-taker is either master or nonmaster. A reliable and valid test will lead to a judgment that matches the true status with the test-taker's performance. If the test-taker is a nonmaster and is classified as such, or the test-taker is a master and is classified as a master, then we have made the correct decisions. However, if the master is judged to be a nonmaster, we have made an error of rejection (also called a false negative because the negative decision on mastery is an error). If a nonmaster is judged to be a master, then we have made an error of acceptance (also called a false positive because the positive decision on mastery is an error). Figure 9.1 summarizes these relationships. The only way to minimize these errors is to ensure that your test is reliable and valid.

THE NECESSITY OF HUMAN JUDGMENT IN SETTING A CUT-OFF SCORE

Several of the techniques we are about to describe have the appearance of statistical precision in establishing a cut-off score. These methods, especially the contrasting groups process, can be appealing because they create the impression of certainty. Don't be deluded. *There is no simple, cookbook solution to establishing the standards for your test, and there is no formula for determining the cut-off score that eliminates the sticky business of human judgment in standard-setting procedures!* With this caveat, there are four considerations worth your attention: consequences of misclassification, stakeholders, revisability, and performance data.

Consequences of Misclassification

One of the first judgments you must make has to do with the consequences of misclassification of test-takers. If it is particularly important to prevent nonmasters from being certified as masters,

FIGURE 9.1 Outcomes of a Criterion-Referenced Test

TRUE STATUS	Nonmaster	Master
Master	Error of Rejection	Correct Decision
Nonmaster	Correct Decision	Error of Acceptance
	TEST DECISION	

it makes sense to raise the cut-off score beyond what might be otherwise satisfactory. On the other hand, if greater damage is done by denying master status to those who may in fact be masters, then it might be advisable to lower the cut-off score.

For example, the consequences of passing a nonmaster during a surgical residency far outweigh the consequences of occasionally holding back a master. In other situations, such as minimal competency tests for high school graduates, the opposite may be true. It may be felt that denying diplomas to masters has such severe social and economic consequences for those individuals that these consequences outweigh the disadvantages to society of letting some nonmasters erroneously graduate.

There are specific consequences to the organization of making both false negative and false positive errors. As a result of false negatives, the company may lose the services of a competent performer, or at the least, lower the morale of the employee. If the test-taker is from a legally protected group (see Chapter 12), there can also be legal costs associated with the erroneous classification of a master as a nonmaster. False positives can cost the company the dollar value of the employees' mistakes as well as lost time on the job as incompetent employees learn the skills they need, usually inefficiently. A damaged reputation for the company due to a failure to perform, and thus a consequent loss of sales, may result from certifying nonmasters as masters. Lawsuits from clients are also a possible consequence of false positive errors. Thus one of the first issues a test designer needs to address when setting standards is the consequences of misclassification resulting from the test whose cut-off score is to be determined.

Stakeholders

It is usually advisable for the standard setters to collect opinions from all groups who have a stake in the outcome of the test decision. In fact, this process is so important that it forms the basis for one of the cut-off score techniques we will cover. For a given corporate test, you might expect to find that the test score you develop is of interest to the EEO officer, the test-takers' supervisors, the personnel department, etc. Thus by using a process that recognizes not only your best professional judgment to set the cut-off score, but also involves those who will be affected by the results of the test, you will cultivate greater acceptance of the testing procedure and its resulting decisions.

Revisability

Never assume that the initial cut-off score will remain un-
changed, or that an established cut-off score will not need to be
revised over time. Experience with the test, new data, or contextual
changes over time may require an adjustment in the cut-off score.
For example, a nuclear power system may have chosen one cut-off
score to certify reactor operators. However, after a year, with data
on critical operator errors, a new, higher score might be selected.
In this instance, excessive allegiance to the original cut-off score
may not only hamper efforts to improve the testing process, it may
lead to a fatal error.

Performance Data

Standard setters should rely extensively on performance data in
choosing the cut-off score. It is unwise to set a standard in the
absence of hard data about how real test-takers perform on the test.
An unrealistically high cut-off score can be particularly difficult to
lower for political or legal reasons, for example, "How come I had to
have a 94% to get a pay grade increase and now people only need a
75%?" At best, an unrealistically low standard can create a sense of
mockery for its graduates; at worst, low standards support a haz-
ardous workplace with unskilled employees damaging equipment or
jeopardizing the health and safety of others.

THREE PROCEDURES FOR SETTING
THE CUT-OFF SCORE

With these thoughts in mind, we can now turn to the first of
three different but complementary methods for determining the cut-
off score for a criterion-referenced test. Each approach is based on
a systematic collection and analysis of data from a different source.
The informed judgment method draws primarily on the perceptions
of various stakeholders in the organization. Conjectural methods
base cut-off scores on content experts' projections of competent performance on each test item. The contrasting groups method uses
performance data of masters and nonmasters to establish the level
of mastery. We recommend that you use as many of these methods

as you can to establish the test's cut-off score—the *political* process of informed judgment, the *projected* outcome of the conjectural method, and the *performance* process of contrasting groups. We also want to emphasize again that the chosen cut-off score should not be considered as absolutely final. The operation of the chosen cut-off score should be monitored periodically to make sure that it is rendering decisions that are satisfactory to those involved and facilitating the achievement of the company's objectives.

Informed Judgment Method

The informed judgment method acknowledges most explicitly that the standard-setting process is essentially one of human judgment. The steps of this method follow directly from paying attention to the four considerations we have just discussed.

1. Begin by analyzing the consequences of misclassification. The political and legal, but especially performance consequences of test-takers on the job need to be analyzed.

2. Gather relevant performance data to see how different test-takers actually do on the test. It is preferable to collect performance data from at least three different groups of test-takers:

    • those who have not taken instruction in the competencies covered by the test
    • those who have just finished such instruction
    • those who are doing the job

    The idea here is to get a feel for how well naive test-takers can do on the test, and how long the test competencies are retained once the instruction is completed. These scores are also considered in light of how actual job performers do.

3. The third step is to collect the preferences of other stakeholders. In other words, take the test to those who will be affected by the consequences of your test decisions, for example, the test-takers' supervisors, test-takers' coworkers, the EEO office, etc., and ask them what they think the cut-off score should be.

4. Finally, select the cut-off score based on a consideration of all the accumulated data. There is no formula available to weight each component and then calculate the final cut-off score. The final decision is thus a professional, informed judgment regarding what level of test-taker competence constitutes mastery.

PRACTICE

Assume that a test was developed to assess the competencies to become an instructional evaluation specialist. Here is the data gathered by the test developer to reach an informed judgment on the cut-off score. What would you recommend as the cut-off?

Average score of nonmasters:	34%
Average score of masters	79%
Preferred cut-off scores of various constituencies:	
Current evaluation specialists	70%
Evaluators' supervisors	85%
Instructional design staff	60%
Evaluators' clients	90%
Personnel Department	50%

Cut-off score estimate   ___%

FEEDBACK

Because this is an informed judgment process, there will generally be variation in suggested cut-off scores due to different divisional responsibilities and different types of interactions with the employees the test will certify. The process of gathering the data and discussing the various constituency judgments is critical to the effectiveness of this technique.

The Personnel Department may wish to have a lower score (50%) to reduce its difficulties in recruiting or to protect itself from an EEO challenge (by viewing the risk of certifying a non-master as less than the risk of legal challenge for denying certification). The clients, of course, would prefer to deal only with the very best professionals and would prefer the instructional evaluation specialist to score at the top end (90%). The instructional design staff could feel threatened by the evaluation specialists and might therefore lobby for a lower skill level. The evaluators' supervisors want to have strong colleagues, but probably have a more realistic view of what the meaning of a specific score on this test is, and will opt for a cut-off score of 85%, still higher than most, but lower than that of the clients. Finally, the current evaluation specialists have probably sought a balance between what is an adequate professional assessment and personal interests in not making their work life any more difficult than it currently is. Hence they have chosen a mid-range score of 70%. (Training professionals reviewing this data—as an exercise divorced from a real situation—have tended to set the cut-off score in the 70–75% range.)

A Conjectural Approach, the Angoff-Nedelsky Method

There is a general class of techniques that rely on professional estimates or conjectures of success for determining the cut-off score of a test. This class of techniques, often referred to as conjectural methods, determines estimates of success for a minimally competent performer on each item. Of the techniques that exist, the Angoff adaptation of the Nedelsky method, or Angoff-Nedelsky for short, is perhaps the most useful technique and the one we want to discuss. It consists of three steps.

1. The first step is to identify judges who are familiar with the competencies covered by the test, and with the performance level of masters of these competencies. The number of judges you select will depend on availability of judges, criticality of the performance, etc. However, we think you would rarely need more than five, with three being the more typical number.
2. The judges are then asked to review each item in the test. For each item, each judge estimates the probability that a minimally competent test-taker would get the item right. Make sure the judges understand that a probability level should never be lower than the level of chance predicted by the item, for example, if there are four alternatives in a multiple-choice item, the estimate should not be lower than 25%.

    These estimates are expressed as percentages and assigned a corresponding decimal value. For example, if a judge thinks there is a fifty-fifty chance of the minimally competent test-taker getting a given item right, that item is assigned a value of .50. If the judge estimates that an item is so simple that the minimally competent test-taker will almost surely get it right, then the item would be assigned a value of 1.00.

    If possible, judges should estimate the probability for each item independently, and then discuss among themselves those items where they disagree markedly in their estimates.
3. The chosen cut-off score is the sum of the probability estimates. If more than one judge is used, the cut-off score is the average of the sums.

Here is an example (Table 9.1) of the process illustrated with a five item test.

TABLE 9.1 Judges' Probability Estimates
Angoff-Nedelsky Method

Item	Judge 1 Probability	Judge 2 Probability	Judge 3 Probability
1	.33	.50	.40
2	.80	.90	1.00
3	.20	.33	.20
4	.20	.90	.33
5	.50	.75	.50
Total	2.03	3.38	2.43

Averaging the Totals for Each Judge to Obtain the Cut-Off Score

$$2.03 + 3.38 + 2.43 = 7.84$$
$$\frac{7.84}{3} = 2.61$$

Cut-off Score $= 2.6$

PRACTICE

Assume you are a judge for a test designed to assess competencies to become an instructional evaluation specialist. For each item estimate the probability that the minimally competent test-taker will get the item correct.

Item   *Probability*
1. Nearly every high school graduating class selects a valedictorian and a salutatorian. What kind of decision do these choices represent?
   a. Norm-referenced
   b. Criterion-referenced
   c. Domain-referenced
   d. None of the above   _____
2. Ann goes bowling. Every time she rolls the ball, it goes into the gutter. In testing terms, Ann's performance might best be described as
   a. both reliable and valid
   b. neither reliable nor valid
   c. valid, but not reliable
   d. reliable, but not valid   _____
3. Mr. Kinser is charged with determining the quality of finished cabinets produced by his

employees. Which is the most reliable instrument he could develop for this purpose?
a. A behaviorally anchored rating scale
b. A checklist
c. An objectives based test
d. A descriptive scale   _____
4. The appropriate correlation coefficient for
calculating the correlation between decisions based on a criterion-referenced test and
IQ is the point-biserial correlation coefficient.
a. True
b. False   _____
5. Define the term "validity."

_____

_____

_____   _____

Cut-off Score (sum of estimates)   _____

FEEDBACK

The estimates for each item can only be made as a function of
the difficulty of the content and the relative level of sophistication
of the minimally competent person in a given organization. Hence
the need to employ qualified judges. Table 9.2 shows what might
be one set of estimates for the minimally competent performer:

**TABLE 9.2 Possible Probability Estimates
                 Angoff-Nedelsky Method**

Item	Probability
1	.75
2	.50
3	1.00
4	.75
5	.75
Cut-off Score Estimate	3.75

Contrasting Groups Method

The contrasting groups method has the aura of scientific precision about it. Indeed, it is probably the single strongest technique of the three, and one may be tempted to use it exclusively. However, there is no evading the necessity for human judgment in setting a cut-off score—even in the contrasting groups method—nor should one standard-setting technique be used alone. We feel the contrasting groups process is most appropriately an integral part of the data collected for the informed judgment method.

1. The first step in this technique is to choose judges who are familiar with employee performance and knowledgeable enough about the competencies tested to select a pool of definite masters and nonmasters. It is important that the judges understand the definition of a "nonmaster."

    Nonmasters are *not* those people who are totally ignorant on the topic to be tested. A nonmaster for testing word processing skills with a specific program would be someone who knew how to type and use a computer—not someone who has no skills with either the computer or the typewriter. Because this technique compares the performance of the two groups, defining a nonmaster as someone totally unskilled in the area may lead to an artificially low standard of mastery.

    If there are no other sources of data to choose from, you can approximate master and nonmaster performance by administering a pre-test and a post-test for a course (if one exists that teaches the skills tested). However, this approach is particularly susceptible to error introduced by poor teaching or underestimating entry skills.

2. Have the judges identify masters and nonmasters. There should be at least 15 people in each group, preferably 30 or more. (As we discussed with test length in Chapter 4, more is usually better. The larger your sample, the more confidence you can have in your judgments about the cut-off score.)

3. Administer the test to both groups, i.e., the masters and the nonmasters.

4. Plot the scores of both groups as frequency distributions. Figure 9.2 illustrates one such pair of frequency distributions.

5. Make the initial cut-off score where the two distributions intersect. In Figure 9.2 this point would be around 42%. This intersection point is the cut-off score that minimizes the total number

FIGURE 9.2 Contrasting Groups Method of Cut-off Score Selection

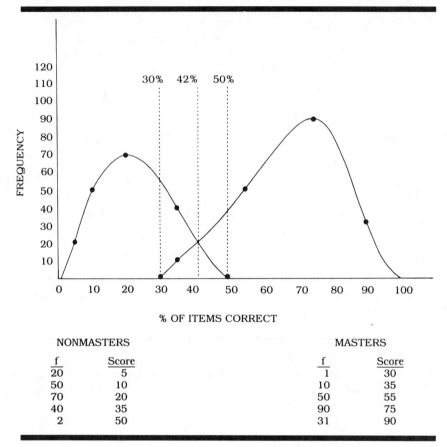

% OF ITEMS CORRECT

NONMASTERS			MASTERS	
f	Score		f	Score
20	5		1	30
50	10		10	35
70	20		50	55
40	35		90	75
2	50		31	90

of misclassifications—both false positive and false negative—resulting from the test.

6. Adjust the cut-off score as necessary. Shifting the cut-off score to 50% would eliminate all nonmasters as well as the lowest scoring masters. Shifting the cut-off score to 30% would create a situation where all identified masters are passed as well as the higher scoring nonmasters. The final cut-off score could be adjusted to any point between 30% and 50%, depending upon the severity of the consequences of false positive and false negative errors and the opinions of other stakeholders.

PRACTICE

A group of masters and nonmasters have been given a test developed to assess the competencies required to become an instructional evaluation specialist. The frequencies of the achieved scores (expressed as percentages of items correct) for both groups are given below in Table 9.3. Plot the frequencies and determine the cut-off score.

TABLE 9.3 Example Test Results for Using the Contrasting Groups Method

Nonmasters		Masters	
Frequency	% Correct	Frequency	% Correct
8	10	1	40
23	20	4	50
30	30	8	60
26	40	17	70
17	50	30	85
8	60	23	90
2	70	1	100

Estimated Cut-off Score _____

FEEDBACK

The graphs below (Figure 9.3) illustrate the results of plotting the two distributions resulting from this test.

As you can see, the initial cut-off score would be set at 60%. By shifting the cut-off score to the right to 75%, you will reduce false positives and increase the chances that only masters will be certified by your test. Raising the cut-off score, however, increases the likelihood of false negative errors. A shift down the scale to 40% will reduce the false negatives, but increase the chances that nonmasters will be certified. The point at which the two curves intersect, i.e., 60%, minimizes the total number of misclassifications (the false positive and false negative misclassifications added together).

Having reviewed these distributions you might go back to the informed judgment exercise and decide whether or not you would have made the same cut-off decision, having seen the distributions rather than just relying on the average scores for masters and nonmasters. When given this fictitious data as a learning exercise, most training professionals tend to shift their cut-off score estimates in light of the distributions—usually to no less than 60% and no more than 75%.

FIGURE 9.3 Feedback, Contrasting Groups Method

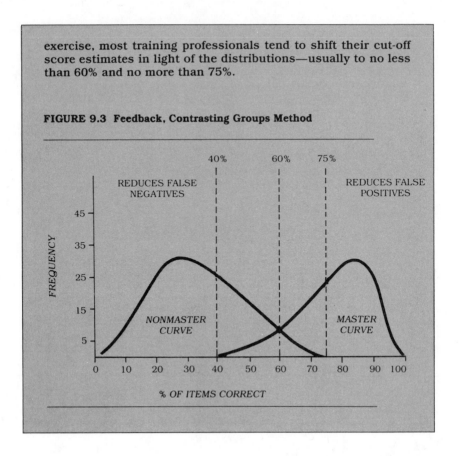

BORDERLINE DECISIONS

As previously mentioned, one of the assumptions of testing has been that any observed score is composed of two parts: true score and error. If you can reduce the error component to zero or near zero, then you can have a high degree of confidence that a person's observed score is equivalent to his or her true score. However, we know that it is very difficult to eliminate the error component of a test score. At the same time, intuitively we understand that at some point people with two different scores really are different in skill level or the knowledge base assessed by the test. For example, most people would agree that two people with scores of 85% and 25% are probably at different levels of mastery. But what about two people with scores of 84% and 86%? Can you really feel confident that they are truly different? What if the cut-off score is set at 85%? Does it make sense to certify one of these test-takers as a master and one as a non-master? Most test designers wonder how to treat the borderline cases, but before we discuss this issue, we need to introduce the concept of standard error of measurement.

The Meaning of Standard Error of Measurement

Beyond intuition, there is a statistical mechanism for estimating the probability that two scores are truly different. The estimate of the amount of variation to be found in a given test score is the standard error of measurement. *In criterion-referenced tests, the standard error of measurement should be considered as a conceptual framework for understanding the quality of the test. Traditional norm-referenced formulas will not be effective with mastery distributions.*

If we had a test of 100 points and one person had a 70 and another a 71, most people (test-takers and designers alike) would not feel comfortable in saying that the person with a 71 really did outperform the person with a 70. If one scored 60 and the other 70, then those involved would probably be more likely to agree there was some difference in performance. For a given score, the standard error provides an estimate of how far apart the scores have to be to be significantly different.

An Example of the Effect of Standard Error. If you knew that a test's reliability was .80 and the standard deviation was .15, by applying a statistics formula you would find that the standard error of measurement is .07 (the ideal standard error of measurement being 0.0). Knowing the standard error of measurement allows you to establish a confidence interval around the observed score. In Figure 9.4 a calculation has been made for someone whose score is 60% on this test, for which you want a level of confidence of 95%. To

FIGURE 9.4 Application of the Standard Error of Measurement

$$
\begin{aligned}
\text{Estimate of true score} &= \text{observed} \pm (\text{Confidence} \times \text{SE}_m) \\
&= .60 \pm (2 \times .07) \\
&= .60 \pm (.14) \\
&= .46 \text{ to } .74
\end{aligned}
$$

FIGURE 9.5 Test Score as a Range, Not a Point

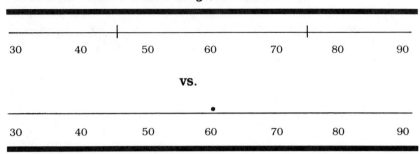

achieve this level of confidence, you will have to establish a range of two standard errors of measurement on either side of the observed score. (If you are willing to accept greater risk of error, you can establish the interval at one standard error of measurement on either side of the observed score.)

In this instance, we would say that there is an approximately 95% probability that this person's observed score of 60% does not deviate from his or her true score by more than 14 percentage points on either side of 60%—by more than two standard errors in either direction. What is most important to remember is that a single test score should always be viewed as a point within a range—not a single, absolute point (see Figure 9.5).

Problems with Standard Error and Mastery Distributions. The standard error of measurement is a useful concept to understand, but its application to most criterion-referenced tests is controversial. The controversy is centered on the argument that it should be applied only to those items that measure a single objective. In other words, many authorities feel that reliability estimates need to be figured on subsets of test items that measure the same objective. This recommendation is usually impractical, since most tests will measure more than one objective, with each objective being assessed by a small number of items. Following this advice would result in a multitude of error estimates, none of them very accurate because they were based on so few items. So what should you do? *If you have any variability in your test scores, and you have access to a measure of internal consistency (see Chapter 11), the standard error of measurement would be a useful guide to the width of the border of indecision surrounding individual scores.* However, there is a practical and intuitive approach to dealing with the borderline cases.

Reducing Classification Errors at the Borderline

If you are particularly concerned about certification errors for test-takers who score near the cut-off score (above *or* below the cut-off score), then you can follow a simple procedure we have adapted from Robert Lathrop (1986) to resolve this problem.

1. Establish a region of indecision, the width of the border, around the cut-off score. Ideally this border would be defined as a function of the standard error of measurement; but an estimate of the width of this region can be made by selecting one tenth of the total test range around the cut-off score. For example, on a test with 100 points, and a cut-off score of 85, the border would be between 80 and 90. (Lathrop favors one fifth of the total test range, but estimates are primarily for norm-referenced tests with greater variability in scores than the criterion-referenced test.) The width of the border is still a matter of professional judgment associated with the test designer's assessment of the risks of misclassification. The greater the concern about misclassification, the larger the border may be.

2. For those test-takers who have scores within the borderlines, you should administer a second testing. With the second test, three outcomes are possible:

    *   The test-taker will score above the cut-off score on both tests.
    *   The test-taker will score below the cut-off score on both tests.
    *   The test-taker will score above the cut-off score on one test and below it on the other.

    When the scores are the same on both tests, i.e., master/master or nonmaster/nonmaster, then the decision is made as indicated by the tests. If the scores are different on the two tests, then you are back to the issue of subjective judgment and the risk level you are willing to accept for misclassification. If the consequences of certifying a nonmaster as a master are great, then you may choose to classify the inconsistent test-taker as a nonmaster. If the consequences of judging a master as a nonmaster are greater, then you might choose to certify the inconsistent test-taker as a master.

In the end, and once again, there is no substitute for professional judgment. However, if you are going to establish a procedure for borderline cases, do so in advance of the test and make this procedure known to all test-takers. Inconsistency in testing procedures, especially if they affect a protected group (see Part V), may well create negative professional, ethical, and legal consequences.

10.

Establishing the Reliability of Performance Tests

RELIABILITY AND VALIDITY OF PERFORMANCE TESTS
INTER-RATER RELIABILITY
REPEATED PERFORMANCE AND CONSECUTIVE SUCCESS
PROCEDURES FOR TRAINING RATERS

RELIABILITY AND VALIDITY OF PERFORMANCE TESTS

The concepts of reliability and validity are as important in performance testing as in paper-and-pencil testing. However, the reliability and validity problems associated with performance testing are different from those associated with paper-and-pencil tests composed of closed-ended items. In some ways the reliability and validity problems associated with performance tests are similar to those posed by essays and other types of open-ended questions; unlike closed-ended assessments that are even machine scorable, the test-taker's behavior on an essay test or during a performance test must be rated or judged by an observer. Therefore, the locus of reliability and validity shifts from the test itself to the consistency of the test-taker's performance and the reliability and validity of the judges' observations.

The reason for the creation of rating scales as described in Chapter 5 is to improve the reliability and validity of these observers' judgments. The more specific the rating instrument, the better it supports reliable and valid observations. However, error in these judgments is always a concern, and test designers should be aware of the types of errors that most frequently occur, how to assess the reliability of the judgments, and how to correct unreliable and invalid performance assessments when they are revealed.

The problem of consistency in the test-taker's performance is in many ways analogous to the reliability issue presented in Chapter 4 of how many items should be on a paper-and-pencil test. The issue is one of adequate sampling of the test-taker's ability.

Curiously enough, validity is usually not a problem with performance tests that have been shown to be reliable. Unlike paper-and-pencil tests, performance tests deal with observable actions rather than indicators of mental processes, and so their validity is usually assured once consistency in the observations of raters has been achieved.

This chapter begins with a discussion of the five most common types of error associated with rating scales: errors of standards, halo, logic, similarity, and central tendency. Two different methods of determining the reliability of the raters' judgments are then explained, followed by a discussion of the role of repeated performance in performance testing. The chapter closes with some procedures for training the raters to increase the reliability of their judgments.

Types of Rating Errors

Error of Standards.   In discussing checklists in Chapter 5, we noted that the scale, in order to be considered valid, must contain descriptions of behaviors or characteristics that accurately reflect the desired performance or product outcome. A failure to define the standards in a precise manner on any scale is an "error of standards." The major flaw with numerical and descriptive scales is their inability to provide definitions of behaviors specific enough to prevent raters from imposing their own interpretations on the standards, and subsequently rating the same performance or product differently.

Halo Error.   The "halo error" is a tendency on the part of the raters to allow, usually quite subconsciously, a performance judgment to be influenced by their own opinion of the performer. Despite its name, the halo error isn't always a positive one. A halo effect can also occur when a person's score on a performance test is negatively affected by a rater's opinion. For example, if an instructor were to serve as a rater to assess an end-of-course performance, and the "best" student in the class performed poorly on the performance test, a halo error would occur if the instructor rated the student's performance higher than it actually merited. Subconsciously, the rater may have been thinking, "She *really* knows this stuff, she just had an off day. . . . no point penalizing her." The opposite sentiment introduces just as much error into the testing process—a particularly difficult student performs well on a test, but is downgraded by the instructor because of a previously formed bias against the test-taker's competence.

Logic Error. A "logic error" occurs when a rater is supposed to be rating one characteristic but is really rating another. This type of error happens when the rater is confused about or unaware of the independence of the characteristics of the performance. For example, suppose an airline refueling supervisor assumes that haste is the cause of unsafe fueling practices when in fact the time used by the technician and the number of safety violations are unrelated. Under these circumstances this supervisor will tend to assume incorrectly that technicians who take longer to perform this task are safer; he or she will rate as safer those who spend more time, instead of focusing specifically on safety procedures such as attaching a grounding strap. The confusion of time with safety is a logic error that is correctable by a task analysis that results in the creation of a valid and specific checklist.

Similarity Error. This error is sometimes called the "similar-to-me error." There is some evidence that raters will tend to rate performers they perceive as similar to themselves more highly than those who are "different." Such frequently irrelevant characteristics as educational background, job experience, sex, race, etc., could thus lead to incorrect assessments of performance in the absence of carefully designed rating instruments and adequate rater training.

Central Tendency Error. There is a distinct pattern of rater behavior that is associated with any rating scale that allows a rater to choose points along a continuum, as with descriptive, numerical, or behaviorally anchored scales. Raters avoid the extremes of the scale. Their tendency is to group ratings in the middle, hence the term "central tendency error." This finding is so consistent that for a given scale, the two extreme positions will likely be lost; thus, a seven-point rating scale becomes really only a five-point scale. Perhaps the thinking is, for example, "Nobody is really perfect. If I rate them as that good, people will wonder about me. . . . Better not stick my neck out."

When you are constructing rating scales, give thought to using only an even number of categories, thus removing the exact center point and forcing more spread in the ratings. (Exceptional scales are those where the end points define opposite and equally undesirable characteristics, such as too hot and too cold or too tight and too loose; the meaning of this type of scale requires a center point.) Another strategy is to use a larger number of points on the scale (say, eight rather than five), assuming that raters will tend not to use the extremes.

In the end, it is best to use a checklist or a behaviorally anchored scale. Error is reduced primarily by the precise specification of criteria. The precision of behavioral specification is as important in testing as it is in the design of instruction.

INTER-RATER RELIABILITY

It is important to establish the consistency among different raters' judgments of the performances of test-takers. The mastery/non-mastery decision made about each test-taker should be determined by what the test-taker does, not by differences among the judges, either in what they see or in the value they place upon what they see. Remember that reliability is a prerequisite for validity. Therefore, if there is no inter-rater reliability, i.e., if the judges are inconsistent, their decisions cannot possibly be valid.

The two methods we shall demonstrate for assessing inter-rater reliability yield comparable results, but are conceptually different. One is based upon a corrected percentage of agreement figure (kappa or κ), while the other is a correlation coefficient (phi or ϕ). Which one you decide to use really depends on which of the two statistics you find the easiest to understand or perhaps which of the two you think other interested parties will find the easiest to understand. Persons who have worked with correlation coefficients in the past will probably find phi conceptually more appealing. Those totally unfamiliar with correlation and the meaning of its range of values will probably prefer to use kappa.

These statistics are the same ones we use to determine test-retest reliability in Chapter 11, where their interpretation is explained in detail. In this chapter they will be modified in order to be applicable to more than two judges. However, if the agreement coefficent or the kappa coefficient is unfamiliar to you, you should read the sections on Description of the Agreement Coefficient (p.177) and Description of Kappa (p.179) in Chapter 11 before calculating kappa. Likewise, if the concept of correlation or the phi coefficient is unfamiliar to you, you should read the sections on Correlation (pp.164) and Description of Phi (pp.172) in Chapter 11 before calculating phi.

Calculating and Interpreting Kappa

The kappa coefficient (κ) was designed to measure the agreement between two judges; however, averaging procedures allow you to calculate kappa for more than two judges. This is an important point because we strongly recommend that you have more than two judges when trying to establish the reliability of your performance testing procedures. We will begin by showing how to calculate kappa for two judges and then extend the procedure to three. You will see how you can use the same process to extend the procedure to any number of judges.

The formula for calculating kappa is

$$\kappa = \frac{p_o - p_{chance}}{1 - p_{chance}}$$

Let's begin by explaining what this formula means. The calculation of kappa begins by figuring the percentage of test-takers consistently classified by two judges. This number, called the agreement coefficient (p_o), is inflated by chance agreements. In other words, this number will give you a false sense of security in the reliability of your judges. So this percentage of observed agreement (p_o) is corrected for these chance agreements by subtracting the number of agreements that would be expected due to chance alone (p_{chance}). The result of this subtraction is then divided by $1 - p_{chance}$. The result of this subtraction represents the maximum possible improvement over chance agreement that the two judges could possibly make; so the result of the division represents the proportion of possible improvement in agreement beyond chance agreement actually achieved by the two judges. In this same way kappa is calculated for each pair of judges that you have. The resulting kappa coefficients are then averaged to determine the kappa coefficient for your entire panel of judges.

Table 10.1 presents some possible results of a performance test where three judges rated each of 10 test-takers.

The easiest way to do a kappa calculation is to begin with a matrix for each pair of judges to organize your test results. This matrix will provide you with the numbers you need to calculate p_o and p_{chance} and, therefore, to calculate kappa for each given pair of judges. If you have three judges, you will need three matrices, one for each possible pairing of the judges. Therefore, if you have three judges, you will need a matrix for the Judge 1/Judge 2 pair, a matrix for the Judge 1/Judge 3 pair, and a matrix for the Judge 2/Judge 3

**TABLE 10.1 Example Performance Test Data
Inter-Rater Reliability**

Test-Taker #	Judges 1	2	3
1	Master	Master	Master
2	Master	Master	Master
3	Master	Master	Nonmaster
4	Master	Nonmaster	Nonmaster
5	Nonmaster	Nonmaster	Master
6	Nonmaster	Nonmaster	Nonmaster
7	Nonmaster	Nonmaster	Nonmaster
8	Nonmaster	Nonmaster	Nonmaster
9	Nonmaster	Nonmaster	Nonmaster
10	Nonmaster	Nonmaster	Nonmaster

pair. (If you had four judges, you would need six matrices, one for Judges 1 and 2, Judges 1 and 3, Judges 1 and 4, Judges 2 and 3, Judges 2 and 4, and Judges 3 and 4.) Figure 10.1 illustrates such a matrix for Judge 1 and Judge 2 in Table 10.1.

The cells of the matrix are defined as: **a** equals the number of test-takers classified as masters by both Judge 1 and Judge 2; **b** equals the number classified as nonmasters by Judge 1 but as masters by Judge 2; **c** equals the number judged as masters by Judge 1, but classified as nonmasters by Judge 2; and **d** equals the number judged as nonmasters by both judges. The numbers from the data in Table 10.1 for all three judges have been correctly placed in the three matrices in Figures 10.2, 10.3, and 10.4. Figure 10.2 shows

FIGURE 10.1 Matrix for Determining p_o and p_{chance}

	JUDGE 1 Master	Nonmaster	
JUDGE 2 Master	a =	b =	(a + b) =
Nonmaster	c =	d =	(c + d) =
	(a + c) =	(b + d) =	

FIGURE 10.2 Example p_o and p_{chance} Matrix, Judges 1 & 2

	JUDGE 1		
	Master	*Nonmaster*	
JUDGE 2 *Master*	a = 3	b = 0	(a + b) = 3
Nonmaster	c = 1	d = 6	(c + d) = 7
	(a + c) = 4	(b + d) = 6	

FIGURE 10.3 Example p_o and p_{chance} Matrix, Judges 1 & 3

	JUDGE 1		
	Master	*Nonmaster*	
JUDGE 3 *Master*	a = 2	b = 1	(a + b) = 3
Nonmaster	c = 2	d = 5	(c + d) = 7
	(a + c) = 4	(b + d) = 6	

FIGURE 10.4 Example p_o and p_{chance} Matrix, Judges 2 & 3

	JUDGE 2		
	Master	*Nonmaster*	
JUDGE 3 *Master*	a = 2	b = 1	(a + b) = 3
Nonmaster	c = 1	d = 6	(c + d) = 7
	(a + c) = 3	(b + d) = 7	

the data for Judge 1 and Judge 2; Figure 10.3 shows the data for Judge 1 and Judge 3, while Figure 10.4 contains the data for Judge 2 and Judge 3.

After the data has been organized into the appropriate matrices, we can begin to calculate p_o and p_{chance} and then kappa for each pair of judges.

The formula for calculating p_o is

$$p_o = \frac{(\mathbf{a} + \mathbf{d})}{\mathbf{N}}$$

where **a** and **d** are obtained from the matrix and **N** equals the total number of test-takers. Substituting the data for Judge 1 and Judge 2 from the matrix, Figure 10.2, we calculate p_o as follows:

$$p_o = \frac{(3 + 6)}{10}$$

$$= \frac{9}{10}$$

$$p_o = .90 \text{ for the pair Judge 1/Judge 2}$$

The formula for calculating p_{chance} is

$$p_{chance} = \frac{[(\mathbf{a}+\mathbf{b})(\mathbf{a}+\mathbf{c})] + [(\mathbf{c}+\mathbf{d})(\mathbf{b}+\mathbf{d})]}{\mathbf{N^2}}$$

where **a, b, c,** and **d** are obtained from the matrix and $\mathbf{N^2}$ is the number of test-takers squared. Once again substituting the data from Figure 10.2, we calculate p_{chance} for Judge 1 and Judge 2 as follows:

$$p_{chance} = \frac{[(3+0)(3+1)] + [(1+6)(0+6)]}{10^2}$$

$$= \frac{[(3)(4)] + [(7)(6)]}{100}$$

$$= \frac{(12 + 42)}{100}$$

$$= \frac{54}{100}$$

$$p_{chance} = .54 \text{ for the pair Judge 1/Judge 2}$$

As indicated earlier, the formula for calculating kappa is

$$\kappa = \frac{p_o - p_{chance}}{1 - p_{chance}}$$

Therefore, substituting the values for p_o and p_{chance} for the Judge 1/Judge 2 pair determined above, we calculate kappa for this pair as follows:

$$\kappa = \frac{.90 - .54}{1 - .54}$$

$$= \frac{.36}{.46}$$

$$\kappa = .7826 \text{ or } .78 \text{ for the pair Judge 1/Judge 2}$$

The data for calculating p_o and p_{chance} for the pair Judge 1/Judge 3 is obtained from the matrix in Figure 10.3. These calculations are illustrated below.

$$p_o = \frac{(2 + 5)}{10}$$

$$= \frac{7}{10}$$

$$p_o = .70 \text{ for the pair Judge 1/Judge 3}$$

$$p_{chance} = \frac{[(2+1)(2+2)] + [(2+5)(1+5)]}{10^2}$$

$$= \frac{[(3)(4)] + [(7)(6)]}{100}$$

$$= \frac{(12 + 42)}{100}$$

$$= \frac{54}{100}$$

$$p_{chance} = .54 \text{ for the pair Judge 1/Judge 3}$$

Having the p_o and p_{chance} values for this pair allows us to calculate kappa for Judge 1 and Judge 3 as follows:

$$\kappa = \frac{.70 - .54}{1 - .54}$$

$$= \frac{.16}{.46}$$

$$\kappa = .34782 \text{ or } .35 \text{ for the pair Judge 1/Judge 3}$$

The corresponding calculations for Judge 2 and Judge 3 are as follows:

$$p_o = \frac{(2 + 6)}{10}$$

$$= \frac{8}{10}$$

$$p_o = .80 \text{ for the pair Judge 2/Judge 3}$$

$$p_{chance} = \frac{[(2+1)(2+1)] + [(1+6)(1+6)]}{10^2}$$

$$= \frac{[(3)(3)] + [(7)(7)]}{100}$$

$$= \frac{(9 + 49)}{100}$$

$$= \frac{58}{100}$$

$$p_{chance} = .58 \text{ for the pair Judge 2/Judge 3}$$

$$\kappa = \frac{.80 - .58}{1 - .58}$$

$$= \frac{.22}{.42}$$

$$\kappa = .5238 \text{ or } .52 \text{ for the pair Judge 2/Judge 3}$$

Now we have three kappa coefficients, one for each pair of judges. In order to obtain the overall kappa coefficient, the symbol for which is $\bar{\kappa}_2$ (Conger, 1980), we must average the three pairwise kappas. This calculation is shown below.

$$\text{kappa for Judge 1/Judge 2} = .78$$

$$\text{kappa for Judge 1/Judge 3} = .35$$

$$\text{kappa for Judge 2/Judge 3} = .52$$

Therefore, the average kappa is

$$\bar{\kappa}_2 = \frac{.78 + .35 + .52}{3}$$

$$= \frac{1.65}{3}$$

$$\bar{\kappa}_2 = .55, \text{ the kappa coefficent for all three judges}$$

This average kappa is interpreted as the average improvement over chance agreement resulting from using the rating instrument and the trained judges. The premise behind the statistic is that if you flipped a coin instead of using trained judges armed with a rating instrument, you would get a predictable amount of agreement between the coin flips; kappa represents the improvement over this chance agreement resulting from using the trained judges and the rating instrument.

It is important to remember that the average kappa does not represent the average percentage of agreement between the judges;

that figure would be obtained by averaging the three p_o figures for the three pairs rather than averaging the three kappas. Kappa is a superior measure, however, because the average percentage of agreement would be badly inflated due to chance agreements alone. Organizations that base decisions about reliability on percentage of agreement figures alone typically don't realize that many times their raters are making decisions about test-takers at levels no better than chance. Employees certainly deserve more consistency in their evaluation than can be obtained by simply flipping a coin to see whether or not they pass a performance test. It is even quite possible for kappa to assume negative values, meaning that raters are operating at levels below chance agreement with one another. Such a circumstance indicates a very serious reliability and validity problem with the instrument or with the judges.

The question of how high a kappa coefficient should be is very difficult. It is so because how reliable the judges have to be depends entirely on the organizational and personal consequences of their being unreliable. As described in Chapter 4 in the section Criticality of Decisions and Test Length, an organization should undertake a systematic examination of what the consequences are to the organization and to the test-takers of making mistakes in master/nonmaster decisions. We consider an average kappa coefficient of .60 a minimum. As you can readily imagine, the value should be higher as the criticality of the performance test increases. Experience in using this statistic will help an organization determine how high the value of kappa should be; after several applications of the statistic, you will get a "feel" for what its various levels mean in terms of levels of agreement between judges.

PRACTICE

Table 10.2 on page 144 provides a sample of possible results from a performance test where three judges rated each of 12 test-takers. Use the formulas above to calculate the average kappa coefficent, $\bar{\kappa}_2$, for these three judges. Three blank matrices (Figures 10.5, 10.6, and 10.7) are provided to assist you.

**TABLE 10.2 Sample Performance Test Data
Inter-Rater Reliability**

		Judges	
Test-Taker #	*1*	*2*	*3*
1	Master	Master	Master
2	Master	Master	Master
3	Master	Master	Master
4	Master	Master	Master
5	Master	Nonmaster	Master
6	Master	Nonmaster	Nonmaster
7	Nonmaster	Nonmaster	Nonmaster
8	Nonmaster	Nonmaster	Nonmaster
9	Nonmaster	Nonmaster	Nonmaster
10	Nonmaster	Nonmaster	Nonmaster
11	Nonmaster	Master	Master
12	Nonmaster	Nonmaster	Master

FIGURE 10.5 Blank p_o and p_{chance} Matrix, Judges 1 & 2

		JUDGE 1	
		Master	*Nonmaster*
	Master	a =	b =	(a + b) =
JUDGE 2			
	Nonmaster	c =	d =	(c + d) =
		(a + c) =	(b + d) =

FIGURE 10.6 Blank p_o and p_{chance} Matrix, Judges 1 & 3

JUDGE 1

		Master	Nonmaster	
JUDGE 3	Master	a =	b =	(a + b) =
	Nonmaster	c =	d =	(c + d) =
		(a + c) =	(b + d) =	

FIGURE 10.7 Blank p_o and p_{chance} Matrix, Judges 2 & 3

JUDGE 2

		Master	Nonmaster	
JUDGE 3	Master	a =	b =	(a + b) =
	Nonmaster	c =	d =	(c + d) =
		(a + c) =	(b + d) =	

FEEDBACK

The three blank matrices in Figures 10.5, 10.6, and 10.7 should have been completed as in Figures 10.8, 10.9, and 10.10 on page 146 respectively.

FIGURE 10.8 Answer for p_o and p_{chance} Matrix, Judges 1 & 2

		JUDGE 1		
		Master	Nonmaster	
JUDGE 2	Master	a = 4	b = 1	(a + b) = 5
	Nonmaster	c = 2	d = 5	(c + d) = 7
		(a + c) = 6	(b + d) = 6	

FIGURE 10.9 Answer for p_o and p_{chance} Matrix, Judges 1 & 3

		JUDGE 1		
		Master	Nonmaster	
JUDGE 3	Master	a = 5	b = 2	(a + b) = 7
	Nonmaster	c = 1	d = 4	(c + d) = 5
		(a + c) = 6	(b + d) = 6	

FIGURE 10.10 Answer for p_o and p_{chance} Matrix, Judges 2 & 3

		JUDGE 2		
		Master	Nonmaster	
JUDGE 3	Master	a = 5	b = 2	(a + b) = 7
	Nonmaster	c = 0	d = 5	(c + d) = 5
		(a + c) = 5	(b + d) = 7	

The kappa coefficients for each of the three pairs of judges should have been calculated as shown below.

For the pair Judge 1/Judge 2:

$$p_o = \frac{(4 + 5)}{12}$$

$$= \frac{9}{12}$$

$$p_o = .75$$

and

$$p_{chance} = \frac{[(4+1)(4+2)] + [(2+5)(1+5)]}{12^2}$$

$$= \frac{[(5)(6)] + [(7)(6)]}{144}$$

$$= \frac{(30 + 42)}{144}$$

$$= \frac{72}{144}$$

$$p_{chance} = .50$$

so kappa can be calculated as follows:

$$\kappa = \frac{.75 - .50}{1 - .50}$$

$$= \frac{.25}{.50}$$

$$\kappa = .50 \text{ for the pair Judge 1/Judge 2}$$

For the pair Judge 1/Judge 3:

$$p_o = \frac{(5 + 4)}{12}$$

$$= \frac{9}{12}$$

$$p_o = .75$$

and

$$p_{chance} = \frac{[(5+2)(5+1)] + [(1+4)(2+4)]}{12^2}$$

$$= \frac{[(7)(6)] + [(5)(6)]}{144}$$

$$= \frac{(42 + 30)}{144}$$

$$= \frac{72}{144}$$

$$p_{chance} = .50$$

so

$$\kappa = \frac{.75 - .50}{1 - .50}$$

$$= \frac{.25}{.50}$$

$\kappa = .50$ for the pair Judge 1/Judge 3

For the pair Judge 2/Judge 3:

$$p_o = \frac{(5 + 5)}{12}$$

$$= \frac{10}{12}$$

$$p_o = .8333 \text{ or } .83$$

and

$$p_{chance} = \frac{[(5+2)(5+0)] + [(0+5)(2+5)]}{12^2}$$

$$= \frac{[(7)(5)] + [(5)(7)]}{144}$$

$$= \frac{(35 + 35)}{144}$$

$$= \frac{70}{144}$$

$$p_{chance} = .4861 \text{ or } .49$$

so

$$\kappa = \frac{.83 - .49}{1 - .49}$$

$$= \frac{.34}{.51}$$

$\kappa = .6666$ or $.67$ for the pair Judge 2/Judge 3

Averaging the three kappa values gives you the overall kappa coefficient as follows:

$$\bar{\kappa}_2 = \frac{.50 + .50 + .67}{3}$$

$$= \frac{1.67}{3}$$

$\bar{\kappa}_2 = .5566$ or $.56$, the kappa coefficient for all three judges

Calculating and Interpreting Phi

Phi (ϕ) is a correlation coefficient used to determine the relationship between two dichotomous variables, i.e., two variables each of which have only two values. As such it is an appropriate statistic to use for calculating the level of agreement between two judges each of whom have assigned test-takers to master/nonmaster classifications. Here again, we recommend that you use more than two judges to establish the reliability of your performance testing system. Phi can be extended to more than two judges by averaging the phi coefficients obtained for each possible pair of judges in a process analogous to that described for kappa above.

One begins to calculate phi by constructing a matrix to summarize the judgments made by each possible pair of judges. As with kappa above, if you have three judges you will need three matrices, one each for the pair Judge 1/Judge 2, the pair Judge 1/Judge 3, and the pair Judge 2/Judge 3. Figure 10.11 illustrates how the matrix for Judge 1 and Judge 2 would look. Note that the matrix for phi is constructed differently from the matrix used to determine p_o and p_{chance} values for calculating kappa. The matrix for phi must be set up in exactly the way shown in order for the formula to provide the correct result.

FIGURE 10.11 Matrix for Calculating Phi

	JUDGE 1		
	Nonmaster	*Master*	
Master	B =	A =	(A + B) =
Nonmaster	D =	C =	(C + D) =
	(B + D) =	(A + C) =	

JUDGE 2 (left label for the *Master* / *Nonmaster* rows)

The three matrices for calculating the phi coefficient on the data in Table 10.1 would be completed as in Figures 10.12, 10.13, and 10.14 for the pairs Judge 1/Judge 2, Judge 1/Judge 3, and Judge 2/Judge 3 respectively.

FIGURE 10.12 Matrix for Calculating Phi, Judges 1 & 2

	JUDGE 1		
	Nonmaster	*Master*	
Master	B = 0	A = 3	(A + B) = 3
Nonmaster	D = 6	C = 1	(C + D) = 7
	(B + D) = 6	(A + C) = 4	

JUDGE 2 (left label for the *Master* / *Nonmaster* rows)

FIGURE 10.13 Matrix for Calculating Phi, Judges 1 & 3

	JUDGE 1		
	Nonmaster	*Master*	
Master	B = 1	A = 2	(A + B) = 3
JUDGE 3			
Nonmaster	D = 5	C = 2	(C + D) = 7
	(B + D) = 6	(A + C) = 4	

FIGURE 10.14 Matrix for Calculating Phi, Judges 2 & 3

	JUDGE 2		
	Nonmaster	*Master*	
Master	B = 1	A = 2	(A + B) = 3
JUDGE 3			
Nonmaster	D = 6	C = 1	(C + D) = 7
	(B + D) = 7	(A + C) = 3	

The formula for calculating phi is

$$\phi = \frac{(AD) - (BC)}{\sqrt{(A+B)(C+D)(A+C)(B+D)}}$$

The next step in the calculation of phi for the three judges is to calculate phi for each of the possible pairs. These calculations are illustrated below.

For the pair Judge 1/Judge 2:

$$\phi = \frac{[(3)(6)] - [(0)(1)]}{\sqrt{(3)(7)(4)(6)}}$$

$$= \frac{(18 - 0)}{\sqrt{504}}$$

$$= \frac{18}{22.45}$$

$\phi = .8017$ or $.80$ for the pair Judge 1/Judge 2

For the pair Judge 1/Judge 3:

$$\phi = \frac{[(2)(5)] - [(1)(2)]}{\sqrt{(3)(7)(4)(6)}}$$

$$= \frac{(10 - 2)}{\sqrt{504}}$$

$$= \frac{8}{22.45}$$

$\phi = .3563$ or .36 for the pair Judge 1/Judge 3

For the pair Judge 2/Judge 3:

$$\phi = \frac{[(2)(6)] - [(1)(1)]}{\sqrt{(3)(7)(3)(7)}}$$

$$= \frac{(12 - 1)}{\sqrt{441}}$$

$$= \frac{11}{21}$$

$\phi = .5238$ or .52 for the pair Judge 2/Judge 3

To find the phi coefficient for the entire panel of three judges, the three pairwise phi coefficients must be averaged. Unfortunately, the averaging of correlation coefficients presents some problems. Correlation coefficients are not separated from one another by equal metric units as are, for example, the numbers from 1 to 10. In fact, differences between large correlation coefficients are greater than differences between small ones (Guilford & Fruchter, 1978). In other words, the correlation coefficients .95 and .99 are further apart than the correlation coefficients .15 and .19. So, sometimes simply figuring the average of correlation coefficients results in a distorted number.

The most reasonable advice is that if your phi correlation coefficients for your pairs of judges are close to one another in value and are not very large, you can obtain the overall phi (or average phi, $\bar{\phi}$) by calculating a simple average—adding the phi coefficients up and dividing by the number of pairs. However, while you hope that the phi coefficients are similar, indicating agreement among the pairs, you also hope that the phi coefficients are large, indicating agreement between each member of every pair. Therefore, if you are successful in achieving high agreement levels between your judges, you may find the average of your phi coefficients is distorted.

Happily there is a procedure for correcting average correlation coefficients for this distortion. The process is quick and easy. Each phi coefficient is converted to what is called a *Fisher's Z coefficient* using a simple table (Edwards, 1973). Zs can be safely averaged. So you add up the Z coefficients and divide by the number of coefficients. This average Z is then converted back to a correlation coefficient resulting in a corrected average phi coefficient ($\bar{\phi}$ corrected) for your panel of judges.

This process is illustrated using the three phi coefficients just calculated for the data from Table 10.1. For purposes of comparison, we will first calculate the simple average of these coefficients.

The simple average of the phi coefficients resulting from the data in Table 10.1 is

$$\bar{\phi} = \frac{.80 + .36 + .52}{3}$$

$$= \frac{1.68}{3}$$

$$\bar{\phi} = .56$$

The corrected phi average is computed by first converting each phi into a Z coefficient, using Table 10.3 on page 154. Each phi coefficient is located in the column labeled r; the r is a standard statistical notation designating a correlation coefficient. The corresponding Z coefficient is read from the column labeled Z.

Using the table, we determine that

phi of .80 equals Z of 1.099 (for the pair Judge 1/Judge 2)

phi of .36 equals Z of .377 (for the pair Judge 1/Judge 3)

phi of .52 equals Z of .576 (for the pair Judge 2/Judge 3)

Calculating the simple average of the Zs follows.

$$\bar{Z} = \frac{1.099 + .377 + .576}{3}$$

$$= \frac{2.052}{3}$$

$$\bar{Z} = .684$$

Now we return to the table to convert our average Z back into a phi coefficient. We discover that the value .684 does not appear exactly in the table; the table doesn't break Zs down that finely. So we choose the Z value in the table closest to our average Z, which is .685. The phi corresponding to this value is .595. Therefore,

$$\bar{\phi}_{corrected} = .595 \text{ or } .60$$

TABLE 10.3 Conversion Table for φ (r) into Z

r	Z	r	Z	r	Z	r	Z	r	Z
.000	.000	.200	.203	.400	.424	.600	.693	.800	1.099
.005	.005	.205	.208	.405	.430	.605	.701	.805	1.113
.010	.010	.210	.213	.410	.436	.610	.709	.810	1.127
.015	.015	.215	.218	.415	.442	.615	.717	.815	1.142
.020	.020	.220	.224	.420	.448	.620	.725	.820	1.157
.025	.025	.225	.229	.425	.454	.625	.733	.825	1.172
.030	.030	.230	.234	.430	.460	.630	.741	.830	1.188
.035	.035	.235	.239	.435	.466	.635	.750	.835	1.204
.040	.040	.240	.245	.440	.472	.640	.758	.840	1.221
.045	.045	.245	.250	.445	.478	.645	.767	.845	1.238
.050	.050	.250	.255	.450	.485	.650	.775	.850	1.256
.055	.055	.255	.261	.455	.491	.655	.784	.855	1.274
.060	.060	.260	.266	.460	.497	.660	.793	.860	1.293
.065	.065	.265	.271	.465	.504	.665	.802	.865	1.313
.070	.070	.270	.277	.470	.510	.670	.811	.870	1.333
.075	.075	.275	.282	.475	.517	.675	.820	.875	1.354
.080	.080	.280	.288	.480	.523	.680	.829	.880	1.376
.085	.085	.285	.293	.485	.530	.685	.838	.885	1.398
.090	.090	.290	.299	.490	.536	.690	.848	.890	1.422
.095	.095	.295	.304	.495	.543	.695	.858	.895	1.447
.100	.100	.300	.310	.500	.549	.700	.867	.900	1.472
.105	.105	.305	.315	.505	.556	.705	.877	.905	1.499
.110	.110	.310	.321	.510	.563	.710	.887	.910	1.528
.115	.116	.315	.326	.515	.570	.715	.897	.915	1.557
.120	.121	.320	.332	.520	.576	.720	.908	.920	1.589
.125	.126	.325	.337	.525	.583	.725	.918	.925	1.623
.130	.131	.330	.343	.530	.590	.730	.929	.930	1.658
.135	.136	.335	.348	.535	.597	.735	.940	.935	1.697
.140	.141	.340	.354	.540	.604	.740	.950	.940	1.738
.145	.146	.345	.360	.545	.611	.745	.962	.945	1.783
.150	.151	.350	.365	.550	.618	.750	.973	.950	1.832
.155	.156	.355	.371	.555	.626	.755	.984	.955	1.886
.160	.161	.360	.377	.560	.633	.760	.996	.960	1.946
.165	.167	.365	.383	.565	.640	.765	1.008	.965	2.014
.170	.172	.370	.388	.570	.648	.770	1.020	.970	2.092
.175	.177	.375	.394	.575	.655	.775	1.033	.975	2.185
.180	.182	.380	.400	.580	.662	.780	1.045	.980	2.298
.185	.187	.385	.406	.585	.670	.785	1.058	.985	2.443
.190	.192	.390	.412	.590	.678	.790	1.071	.990	2.647
.195	.198	.395	.418	.595	.685	.795	1.085	.995	2.994

Table (titled *Table of Z' Values for r*) from pg. 503 in *Statistical Methods*, Third Edition, copyright © 1973 by Allen L. Edwards, reproduced by permission of Holt, Rinehart and Winston, Inc.

Notice that the corrected phi is larger than the simple average phi. This comparison illustrates the distortion in uncorrected average correlation coefficients.

The answer to the question of how high an average phi correlation coefficient should be is tempered by the same considerations that determine how high an average kappa coefficient should be. How reliable the judges must be depends upon the consequences of their being incorrect in their decisions. In general, a phi below .60 should be considered unacceptable. For critical objectives, phi coefficients above .95 should be expected.

PRACTICE

Using the performance test results appearing in Table 10.2 and the formulas and procedures described above, practice calculating the simple average and the corrected average phi coefficients for this test data. Three blank matrices (Figures 10.15, 10.16, and 10.17) are provided to assist you.

FIGURE 10.15 Blank Matrix for Calculating Phi, Judges 1 & 2

		JUDGE 1		
		Nonmaster	*Master*	
Master		B =	A =	(A + B) =
JUDGE 2				
Nonmaster		D =	C =	(C + D) =
		(B + D) =	(A + C) =	

FIGURE 10.16 Blank Matrix for Calculating Phi, Judges 1 & 3

		JUDGE 1		
		Nonmaster	*Master*	
Master		B =	A =	(A + B) =
JUDGE 3				
Nonmaster		D =	C =	(C + D) =
		(B + D) =	(A + C) =	

FIGURE 10.17 Blank Matrix for Calculating Phi, Judges 2 & 3

		JUDGE 2		
		Nonmaster	*Master*	
	Master	B =	A =	(A + B) =
JUDGE 3				
	Nonmaster	D =	C =	(C + D) =
		(B + D) =	(A + C) =	

FEEDBACK

The three matrices should have been completed as in Figures 10.18, 10.19, and 10.20 respectively.

**FIGURE 10.18 Answer, Matrix for Calculating Phi,
            Judges 1 & 2**

		JUDGE 1		
		Nonmaster	*Master*	
	Master	B = 1	A = 4	(A + B) = 5
JUDGE 2				
	Nonmaster	D = 5	C = 2	(C + D) = 7
		(B + D) = 6	(A + C) = 6	

FIGURE 10.19 Answer, Matrix for Calculating Phi, Judges 1 & 3

	JUDGE 1		
	Nonmaster	*Master*	
Master	B = 2	A = 5	(A + B) = 7
JUDGE 3			
Nonmaster	D = 4	C = 1	(C + D) = 5
	(B + D) = 6	(A + C) = 6	

FIGURE 10.20 Answer, Matrix for Calculating Phi, Judges 2 & 3

	JUDGE 2		
	Nonmaster	*Master*	
Master	B = 2	A = 5	(A + B) = 7
JUDGE 3			
Nonmaster	D = 5	C = 0	(C + D) = 5
	(B + D) = 7	(A + C) = 5	

The phi calculation for the pair Judge 1/Judge 2 is

$$\phi = \frac{[(4)(5)] - [(1)(2)]}{\sqrt{(5)(7)(6)(6)}}$$

$$= \frac{(20 - 2)}{\sqrt{1260}}$$

$$= \frac{18}{35.5}$$

ϕ = .5070 or .51 for the pair Judge 1/Judge 2

The phi calculation for the pair Judge 1/Judge 3 is

$$\phi = \frac{[(5)(4)] - [(2)(1)]}{\sqrt{(7)(5)(6)(6)}}$$

$$= \frac{(20 - 2)}{\sqrt{1260}}$$

$$= \frac{18}{35.5}$$

ϕ = .5070 or .51 for the pair Judge 1/Judge 3

The phi calculation for the pair Judge 2/Judge 3 is

$$\phi = \frac{[(5)(5)] - [(2)(0)]}{\sqrt{(7)(5)(5)(7)}}$$

$$= \frac{(25 - 0)}{\sqrt{1225}}$$

$$= \frac{25}{35}$$

ϕ = .7143 or .71 for the pair Judge 2/Judge 3

The simple average of these phi coefficients is

$$\bar{\phi} = \frac{.51 + .51 + .71}{3}$$

$$= \frac{1.73}{3}$$

$$\bar{\phi} = .5767 \text{ or } .58$$

The corrected average phi for the three judges is calculated by first transforming the phi coefficients to Zs:

phi of .51 equals Z of .563 (for the pair Judge 1/Judge 2)

phi of .51 equals Z of .563 (for the pair Judge 1/Judge 3)

phi of .71 equals Z of .887 (for the pair Judge 2/Judge 3)

The simple average of the Zs is

$$\bar{Z} = \frac{.563 + .563 + .887}{3}$$

$$= \frac{2.013}{3}$$

$$\bar{Z} = .671$$

Converting the average Z back into a phi, you should have found that Z of .671 has a corresponding phi value of .585; therefore,

$$\bar{\phi}_{corrected} = .585 \text{ or } .59$$

Notice that because the original phi values were not terribly high nor terribly different, the difference between the simple average phi and the corrected phi is extremely small.

REPEATED PERFORMANCE AND CONSECUTIVE SUCCESS

As discussed in Chapter 4, one issue in creating paper-and-pencil tests is how many items should appear on a test. The point was made that the more items included on the test, the more reliable the test would be. This is so because test items are, in effect, like samples of behavior; the more samples of test-taker behavior you examine, the more accurate your assessment of his or her competence will be. A parallel issue exists in performance testing, though it is not so frequently discussed. That issue is, "How many times should a test-taker be asked to demonstrate a behavior on a performance test before we can make an accurate assessment of his or her mastery of the task?"

The danger in making a mastery/nonmastery decision on the basis of a single performance trial is that some tasks can be performed correctly due to chance. Some skills are so essential and the consequences of error in making the master/nonmaster decision are so severe, that repeated performance trials with consecutive successes are warranted. Robert Lathrop (1983) described in detail a process for determining how many performance demonstrations are required to make a master/nonmaster decision at a given criterion level (for example, mastery = 80% or mastery = 70%) with a prespecified level of confidence. His work makes clear several important (and sobering) points for performance test designers to remember.

- The higher you set the criterion for mastery status, the more quickly you can identify nonmasters, but the greater the number of performance trials required to establish mastery. For example, if the criterion for mastery is performing correctly 80% of the steps in a procedure, you will be able to detect nonmasters in fewer trials than if the criterion were 70%; however, more performance trials are required to establish mastery at a criterion of 80% than at a criterion of 70%.

- The lower you set the criterion for mastery status, the more quickly you can identify masters, but the greater the number of trials required to establish nonmastery. For example, if the criterion for mastery is performing correctly 80% of the steps in a procedure, you will be able to identify masters in fewer trials than if the criterion were 90%; however, more performance trials will be required to establish nonmastery at a criterion of 80% than at a criterion of 90%.

- The greater the precision you desire in making master/nonmaster decisions, the more trials will be required to establish either mastery or nonmastery. It is possible to establish independently the levels of error you can tolerate in false positive (erroneously classifying a nonmaster as a master) and false negative (erroneously classifying a master as a nonmaster) errors. Willingness to tolerate fewer mistakes of either kind will increase the number of performance trials required.

- Statistically speaking, given the high criterion levels typical of corporate performance tests and typically acceptable error rates, it is impossible to classify an individual as a master or a nonmaster on the basis of a single performance trial. It may be possible to establish nonmastery in only two trials, but it typically requires at least four or five performance trials to establish mastery.

- Consecutive success is extremely important in the performance testing of critical skills. The effect of a single failed attempt to perform the task is a dramatic increase in the number of additional successful performance trials required to establish the mastery of the test-taker.

The advice that follows from statistical arguments like Lathrop's is below:

- Be sensitive to the possibility (even the likelihood) of error in performance testing; scrutinize tasks with the intention of identifying those likely to be performed correctly by chance. Such tasks are candidates for repeated performance trials.

- Insist on two or more consecutive successful demonstrations of tasks required for the health and safety of clients or employees and tasks essential for organizational survival.

- Remember that while lowering the criterion for mastery makes it easier to identify masters, the criterion for mastery should rightfully be determined by what level of competence is required to do the job. Changing such a criterion, or cut-off score, should never be taken lightly (see Chapter 9).

- Be alert for tasks that must be performed under a variety of different conditions. If these different conditions influence the likelihood of the task's being performed correctly, it is advisable that the test-taker be asked to demonstrate his or her competence under all conditions essential to successful completion of the task on the job.

- Be especially careful of broadly stated, critical objectives such as, "Given appropriate warning indicators in a simulated nuclear power plant, identify the source of radiation leakage." If on a performance test for this objective, the test-taker is confronted with only a single warning indicator pattern, how do the test designers know that he or she can respond correctly to different patterns? Such objectives require a variety of task demonstrations in order to be assessed with confidence.

- For performance tests of critical tasks, be certain that you have formally established the reliability of your rating instruments and your raters and based the performance test on a thorough and accurate job/task analysis.

PROCEDURES FOR TRAINING RATERS

If you think back to a recent Olympics gymnastics or diving competition, you probably had three observations about the rating process: (1) The judges were very consistent in rating the performances of the athletes; (2) your assessments of an athlete's performance ("Easily a 9.5," "Blew that one, probably a 9.1") were probably close to the judges'; and (3) if one judge was consistently different from the other judges, you began to wonder about the judge's skill or motivation. These insights provide an important perspective on why you need to train your raters and how to train them.

For professional and legal reasons, raters need to make consistent and accurate ratings. Good Olympic judges have a combination of experience and training that can serve as a model for any rating situation. There are some simple steps you can follow to train your raters to these high standards:

1. Bring together those people who are familiar with the skill or product to be rated and who will later be asked to serve as raters.

2. Plan a rater training session where you will have available a sample of performances or products that are to be rated. In an ideal setting you would have a model case performance (or product) where all the attributes of a correct performance are present, a clear non-example of the performance (or product), and a range of stimuli between these two extremes—perhaps with the most common errors illustrated.

   This training session can be based either on a live performance or on a high fidelity media simulation, for example, a videotape of an assembly process. If the rating is of a product, the product itself should be present.

3. The raters should be presented with the first stimulus, usually the model case performance. All raters then use the checklist to review the performance or product. If the performance is me-

diated, the tape can be stopped and an action discussed. If the performance is live, plan on recording the actions, for example, a tape recording of an air traffic controller's interchange, so that raters can discuss specific behaviors that may not be clear.

4. Provide the next stimulus, often the non-example, and have the raters assess the performance (or product) as they would during an actual testing session. Again, record the activities if they are live.

5. Ratings for each behavior should be tabulated as a percentage of agreement among raters and posted for the group to see. Raters then share their assessments, step by step, with the other raters. Points of contention (low percentages of agreement) should be reviewed on the tape and discussed until all the raters understand the reasoning behind the correct assessment.

6. A new stimulus is presented, ratings tabulated and shared, and then discussed. This cycle is continued until the judges have reached a high degree of consistency.

7. One to three final trials are then presented to the group with a final inter-rater reliability, i.e., the average kappa or average phi, established.

8. Document the final measures of inter-rater reliability and collect the stimulus materials.

This entire process will rarely take more than a day's effort. In one particular experience, we established a high inter-rater level of agreement with 18 judges from around the nation in less than a morning. *Remember, though, you are establishing the reliability of the rater and the rating instrument in tandem. The rating instrument by itself can't be considered reliable in the sense that a paper-and-pencil test can be.* When the raters move into the field to conduct performance assessments, don't assume there is no need for further follow-up. After a period of time, for example, six months, collect copies of the ratings where multiple judges have observed the same stimulus and calculate the inter-rater reliability coefficient. If your levels of consistency have dropped below the allowable range, then you should bring the raters together for a refresher course.

Finally, when you need to train a new rater, you can use the stimulus materials (if appropriate) that were originally generated for the initial rater training session. After each stimulus you can compare the new rater's judgments to those of the original group until the new rater has reached or exceeded the originally established level of agreement.

11.

Reliability and Validity

THE CONCEPTS OF RELIABILITY, VALIDITY, AND CORRELATION

Reliability and validity are two important characteristics of any kind of test—norm-referenced tests and criterion-referenced tests, paper-and-pencil tests and performance tests. Reliability and validity describe qualities that any good test must possess. They are important factors in defending a test against legal challenge. As indicated in Chapter 1, reliability refers to consistency in the testing results; validity refers to the test's accuracy in measuring what it is intended to measure. While it is possible for a test to be reliable but invalid, it is not possible for a test to be valid if it is unreliable. An unreliable test doesn't measure anything—at least not the same thing every time it is taken, so it cannot possibly be a valid test of any competency.

While the concepts of reliability and validity are fundamental, there are, in fact, several different types of reliability and validity. A test's reliability and frequently its validity are expressed as numbers, i.e., as reliability and validity coefficients. These coefficients differ depending on the type of reliability and validity being expressed. The concepts of reliability and validity are closely related to the concept of correlation, and most reliability and validity coefficients are correlation coefficients of some kind. Therefore, we begin this chapter with a discussion of the meaning of correlation. Since reliability is a logical prerequisite to validity, we next take up the topic of reliability as the concept is most often applied in paper-and-pencil criterion-referenced tests. (The concept of reliability as applied to performance tests is called *inter-rater reliability.* Procedures for establishing inter-rater reliability were presented in Chapter 10.) We conclude this chapter with a discussion of validity in criterion-referenced testing.

Correlation

Correlation is simply a statistical procedure to determine the relationship between two or more variables. A positive correlation means that as one variable changes, the other changes in the same way. A negative correlation means that as one variable changes, the other changes in the opposite way. A zero correlation means there is no relationship between the variables. Figure 11.1 illustrates how these relationships would look if they were graphed.

The top illustration is an example of two variables that are positively correlated. These variables could be high school grade point average and college board scores. As you can see, when the grade point averages (Variable 1) go up, the college board scores (Variable 2) also increase. In the next graph we have a situation where the relationship is a negative one. This could represent the relationship between the average monthly temperature of a city (Variable 1) and the average monthly snowfall in the city (Variable 2). So, when the average temperature is high, say 60°, the average snowfall will be low, maybe near 0″. Conversely, when the average temperature is low, say 20°, the average snowfall might be 90″—with the points between falling on the line (or close to the line). The third graph illustrates a zero correlation. There is no discernible pattern to the relationship. This might be an example of plotting the variables of height and intelligence, two variables known to be totally uncorrelated.

Correlations are derived by plugging the corresponding values of the two variables to be correlated into a mathematical formula. The result is a number that expresses the strength and direction (positive or negative) of the correlation. The number will always be between −1.00 and +1.00. Correlations of, say, +.98 are very high positive correlations; correlations of −.98 are very high negative correlations. Correlations in either direction that are close to zero (−.15 or +.09, for example) are termed low correlations.

Correlations depend partly on variance in the variables being correlated, i.e, correlations are affected by the range of values held

FIGURE 11.1 Graphs Illustrating Correlations of Different Sizes and Directions

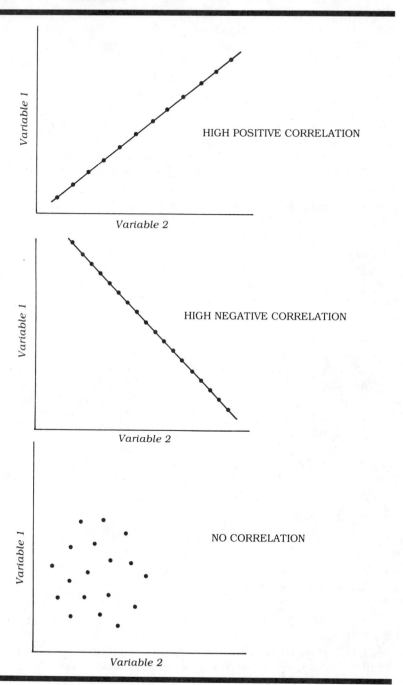

by the variables being correlated. A variable with a small range of values will tend to be only lowly correlated with any other variable. For example, if we were to correlate average monthly rainfall and temperature in a city that is basically the same temperature all year round, we would find the correlation is very low, even if a wide range of rainfall values are reported. This is logically (and mathematically) so because if rainfall varies while temperatures remain the same, there cannot be much of a relationship between them. Correlations between two variables both of which have a small range of values are also low for the same reason. This characteristic of correlations is important for understanding why some of the statistics that work with norm-referenced tests don't work well with criterion-referenced tests. Remember that NRTs "spread the scores of test-takers out" from one another, whereas CRTs are less likely to do so. Hence different correlation coefficients—different reliability and validity indices—are appropriate for the two kinds of tests.

TYPES OF RELIABILITY

It is frequently stated that test reliability refers to the consistency of test scores. In fact, there are several different kinds of reliability—several different ways in which test scores can be consistent. Some are more relevant to criterion-referenced tests in particular than are others. Each of these reliabilities is calculated differently, but most reliability coefficients are correlation coefficients of some type.

There are at least three different kinds of reliability that you might encounter in your work with tests. (There is a fourth—domain score estimation indices, but you are not likely to see these in corporate testing circles.) The three reliabilities we will discuss here are

- internal consistency
- test-retest score consistency
- test-retest mastery classification consistency

Internal Consistency

This measure of reliability determines the extent to which a test measures one underlying ability or competency, i.e., the extent to which the test is internally consistent in what it assesses. Tests with high internal consistency are composed of items that result in the same patterns of responses among test-takers. The most common coefficients of internal consistency are the Kuder-Richardson 20 (K-R 20), Kuder-Richardson 21 (K-R 21), and Cronbach's *alpha* indices. These indices are very likely to appear on any computer-generated item analysis output (of the kind discussed in Chapter 8) or the output of any computer statistical package that calculates test evaluation data. Most of these item analysis and evaluation packages were designed for the evaluation of norm-referenced tests.

There are decided problems in using internal consistency coefficients with the test score results of criterion-referenced tests. These problems arise from two sources:

• Lack of spread (variance) in scores on CRTs
• The inclusion of items that measure unrelated objectives on CRTs

Unlike the results of norm-referenced tests that are designed to separate test-takers from one another, very often there is not much of a range in the test scores resulting from the administration of a criterion-referenced test. The technical term for this situation is a *lack of variance* in the test scores. (As was pointed out in Chapter 2, it is this lack of variance that gives frequency distributions of CRT results their typical tall, narrow shape, in contrast with the bell-shaped distribution resulting from NRTs.) This situation arises because criterion-referenced tests are designed to measure specific competencies, and these competencies have usually been taught to the test-takers in anticipation of the test. Therefore, many of the test-takers will do well on the test, causing the range of scores to be small. Correlations between variables that lack a range of values— that lack variance—will tend to be low. Therefore, internal consistency measures of reliability when applied to a criterion-referenced test will tend to make the test look unreliable.

Another reason why internal consistency measures of reliability are inappropriate for some criterion-referenced tests is that these tests are usually not designed to measure a single, underlying ability anyway, especially in a corporate context. CRTs are composed of items that measure specific objectives; these objectives may or may not be related to one another. If they are unrelated, the items that measure them will not result in similar patterns of responses among test-takers. Since it is the relatedness of the items that internal consistency indices measure, it is questionable whether or not there is much meaning in them for many criterion-referenced tests.

You might ask why we have discussed these indices of internal consistency since their application to CRTs is problematic. We do

so because a little knowledge is a dangerous thing. It is not unusual for someone in a corporate training department to have access to item analysis programs that routinely calculate and output these indices. Sometimes these indices are simply labeled "reliability coefficient." It is important for test developers to know that low internal consistency figures do not necessarily mean that their criterion-referenced test is unreliable. These figures are best interpreted with caution from a position of knowledge regarding what they mean.

Test-Retest Score Consistency

The alternative to internal consistency measures of reliability is what are termed test-retest reliability measures. As the name implies, these measures indicate the consistency of the test scores over time. Their calculation requires two test administrations, either of the same test or of parallel forms of the same test. (Parallel forms are constructed by carefully matching items on the two forms for objective assessed, difficulty level, and other response characteristics discussed in Chapter 8 under item analysis.)

It is desirable that the two administrations be fairly close together in time—between two and five days apart. If too much time elapses, test-takers are likely to have acquired additional relevant information, forgotten information, or otherwise changed in ways that will cause their scores on the second administration to be different. It is important that test-takers receive no additional instruction pertinent to the objectives the test measures during the time between the two administrations.

Test-retest reliability is calculated by correlating the scores from the first test administration with those from the second. Tests with high test-retest reliability will result in test-takers achieving nearly the same scores on both test administrations. Therefore, the resulting reliability coefficient will be a high positive correlation. Professionally developed, standardized, norm-referenced tests will have reliabilities above +.95. The correlation of scores from two administrations of a criterion-referenced test will usually be considerably lower because of the lack-of-variance problem mentioned above. Remember that correlations between variables with a small range—in this case the two sets of test scores—will always tend to be low. Therefore, designers of CRTs should be cautious about drawing reliability conclusions based on a simple correlation of scores from two administrations of the test.

Test-Retest Mastery Classification Consistency

Modifying the test-retest score consistency notion results in a more useful concept of reliability for criterion-referenced tests. Reliability for CRTs can be thought of as the consistency over time of the master/nonmaster decisions that are made based on the test, rather than the consistency of the scores themselves. Here again, two test administrations are required, and all the precautions regarding elapsed time described above apply. With this modification, tests with high test-retest reliability are tests that result in test-takers being classified consistently as masters or nonmasters on two consecutive test administrations. (In a 1988 article by Michael Subkoviak, a process for estimating test-retest reliability from a single test administration is described. However, the procedure relies on a measure of internal consistency reliability and the calculation is more complex than the processes described herein.)

Table 11.1 illustrates the results of two administrations of a criterion-referenced test. Notice that the resulting classifications are fairly consistent, but that 2 of the 10 test-takers (numbers 5 and 10) were misclassified on one of the administrations.

TABLE 11.1 Example of Test-Retest Data for a CRT

1ST TEST ADMINISTRATION		2ND TEST ADMINISTRATION	
Test-Taker	*Status*	*Test-Taker*	*Status*
1. Diana	Master	1. Diana	Master
2. Sid	Master	2. Sid	Master
3. Tom	Master	3. Tom	Master
4. Mary	Master	4. Mary	Master
5. George	Master	5. George	Nonmaster
6. Kenneth	Nonmaster	6. Kenneth	Nonmaster
7. Doug	Nonmaster	7. Doug	Nonmaster
8. Walter	Nonmaster	8. Walter	Nonmaster
9. Polly	Nonmaster	9. Polly	Nonmaster
10. B.G.	Nonmaster	10. B.G.	Master

THE LOGISTICS OF ESTABLISHING TEST-RETEST RELIABILITY

You will need to plan carefully to establish the test-retest reliability of your CRT. Among the factors you must consider are

- choosing items for your trial test
- giving the same test twice versus giving parallel forms
- the sample of test-takers who will take your test twice
- testing conditions during both test administrations

Choosing Items.   Ideally you will have created a pool of items for each of your objectives in excess of the number you think you will actually need (see Chapter 4). This is a recommended strategy because inevitably some items will not work when actually pilot tested. Flaws in the items such as ambiguities and cues will be revealed when the items are tested. Such items should be revised or eliminated from the pool. (Chapter 7 described procedures for piloting your test; Chapter 8 presented information on item analysis statistics and procedures to assist you in identifying poor items.) It is recommended that you eliminate bad items from your pool and select the best items in appropriate numbers for your test before you attempt to establish its reliability. Otherwise, you may find that you have to conduct the reliability procedures again with a substantially revised test. The test should be assembled using the procedures described in Chapter 6.

The Same Test Versus Parallel Forms.   At the point of piloting the test and establishing its reliability, you will want to decide whether you need only one version of your test or more than one parallel forms. If you are creating parallel forms of your test, the reliability of each form should be established. Some arguments for creating parallel forms are as follows:

- Parallel forms allow for retesting of individuals who score too close to the master/nonmaster cut-off score to be classified with confidence.
- Parallel forms can be important if the security of a test is breached; the loose or circulated form can be destroyed and a parallel form placed into immediate service.
- Parallel forms are helpful in case an employee scheduled for group testing has to cancel and take the test at a later date; such an

employee can be given a parallel form of the test without fear that the answers to the test may have been shared.

It is recommended that parallel forms of a test be created by very careful matching of items in terms of the objectives they cover and the ways in which test-takers respond to them, i.e., matched items should have the same difficulty level and the same discrimination index or point-biserial correlation. The difficulty index and the point-biserial correlation are item analysis indices covered in Chapter 8.

Sample Test-Takers. You will need a group of at least 30 people who can take your test twice. Many organizations have difficulty getting a group of that size to take the test. Our feeling is that any test-retest data are better than nothing, so use the largest group that you can arrange. However, be aware that statisticians recommend larger groups for good reasons. The data from small groups are likely to be misleading. With a small group, the performance of even one person can dramatically change your results. Small groups just do not give you the accurate picture of your test's reliability and your items' performance that a larger group will. Our advice is to make every effort to locate a group of 30 people.

The sample of test-takers should be representative of the persons who would take the test under implementation conditions with one additional qualification: Try to ensure that your group includes some masters and some nonmasters; you want to see if your test can reliably distinguish between these two groups. Compose the sample of the types of persons between whom you want the test to distinguish when it is implemented in your organization. Many statisticians would recommend that you draw the sample of test-takers randomly from identified pools of appropriate employees. Most organizations do not have the scheduling flexibility to accommodate this advice. In our opinion it is better to forgo the randomness of the sample than to settle for a sample that is too small.

For documentation purposes you should keep a record of the composition of the test-retest sample (names, titles, or job classifications, etc.) and how the sample was chosen. Such information could be important if the reliability of the test is ever challenged.

Testing Conditions. The two test administrations should be close together, between two and five days. As previously mentioned, it is essential that the sample test-takers not study or receive additional information regarding the content of the test between the two testings. Conditions at each of the two test administrations should be as identical as possible. The test should be given at the same time of day under the same physical, environmental, and psychological circumstances.

CALCULATING RELIABILITY

The reliability of a test is usually expressed as a number called the reliability coefficient. The different forms of reliability described above all have distinct procedures for calculating a reliability coefficient associated with them. We recommended above that you view the reliability of a CRT as the test-retest consistency in master/nonmaster classifications. We follow in this section by demonstrating three different ways to calculate reliability coefficients that meet three conditions: (a) applicability to criterion-referenced tests, (b) relative ease of computation, and (c) relative ease of interpretation. The three methods of calculating coefficients based upon this concept of CRT reliability are

- the phi coefficient (ϕ)
- the agreement coefficient (p_o)
- the kappa coefficient (κ)

The Phi Coefficient

Description of Phi.   Phi (ϕ) is a correlation coefficient that indicates the relationship between two dichotomous variables. Dichotomous variables are simply variables that have only two values, such as male/female, plant/animal, pass/fail, etc. As you can see by examining Table 11.1, criterion-referenced test results constitute such a variable because the status decisions resulting from the test are dichotomous—master/nonmaster. Therefore, one measure of test-retest reliability for CRTs is the phi correlation coefficient calculated on the master/nonmaster decisions from two consecutive administrations of the test.

Like all correlation coefficients, phi can have values from -1.00 to $+1.00$. A phi coefficient of $+1.00$ would indicate that the master/nonmaster classifications were perfectly consistent between the two test administrations, a highly desirable result. A phi coefficient of -1.00 would indicate that everyone who was a master on the first administration was a nonmaster on the second and *vice versa!* This, of course, is a very undesirable and unlikely outcome. A phi coefficient near zero indicates no relationship between the scores on the two administrations. This is also an undesirable outcome, but not unlikely if the test is unreliable. Note that an unreliable test has a phi coefficient near zero, not a high negative correlation near -1.00. A correlation near -1.00 indicates a very strong relationship between two variables; the relationship just happens to be in the opposite direction. Lack of reliability means no, or a weak relationship between the two sets of test scores.

Remember that correlation coefficients are low when one or both of the correlated variables have a small range of values; if all of your sample test-takers are either masters or nonmasters on either test administration, your phi reliability coefficient will be zero.

Calculating the Phi Coefficient. Now let's look at how phi is calculated. The arithmetic is extremely simple and can be done on an inexpensive calculator. One begins by putting the results of the test administrations into a table—a two-by-two matrix, as in Figure 11.2.

In Figure 11.2, cell **B** should contain the number of test-takers who were nonmasters on the first administration but masters on the second. Cell **A** should contain the number who were masters on both test administrations. Cell **D** should contain the number who were classified as nonmasters on both test administrations, while cell **C** should contain the number of test-takers who were masters on the first administration but nonmasters on the second. Check your accuracy; adding the numbers in all four cells should equal the total number of test-takers in your sample. **NOTE:** Your test results must be placed in the table in exactly this way; reversing the posi-

FIGURE 11.2 Phi Table for Test-Retest Reliability

		1ST TEST ADMINISTRATION		
		Nonmaster	*Master*	
2ND TEST ADMINISTRATION	*Master*	B =	A =	A + B =
	Nonmaster	D =	C =	C + D =
		B + D =	A + C =	

tions of the letters in the matrix can result in your phi coefficient appearing to be in the opposite direction from what it actually is.

Once these totals have been placed carefully in these cells, add the numbers in the cells horizontally and vertically as indicated in Figure 11.2. The numbers in the four cells (**B, A, D,** and **C**) and the four cell totals (**A+B, C + D, A + C,** and **B + D**) must then be placed in the following formula, which is the equation for calculating phi.

$$\phi = \frac{(AD) - (BC)}{\sqrt{(A + B)(C + D)(A + C)(B + D)}}$$

Figure 11.3 shows an example of a phi calculation using the data from Table 11.1 presented earlier. First the appropriate totals are placed into the matrix; then the numbers are moved into the formula. The product of **B** times **C** is subtracted from the product of **A** times **D** for the numerator. The denominator is simply the square root of **A + B** mutiplied by **C + D** multiplied by **A + C** multiplied by **B + D,** as illustrated below.

FIGURE 11.3 Example Phi Table for Test-Retest Reliability

		1ST TEST ADMINISTRATION		
		Nonmaster	*Master*	
2ND TEST ADMINISTRATION	*Master*	B = 1	A = 4	A + B = 5
	Nonmaster	D = 4	C = 1	C + D = 5
		B + D = 5	A + C = 5	

$$\phi = \frac{[(4)(4)] - [(1)(1)]}{\sqrt{(5)(5)(5)(5)}}$$

$$= \frac{(16 - 1)}{\sqrt{625}}$$

$$= \frac{15}{25}$$

$$\phi = .60$$

As you can see, the resulting phi coefficient is .60. This indicates that there is a positive correlation between the two sets of test classifications of +.60.

How High Should the Phi Coefficient Be? It seems appropriate to consider at this point how high a phi coefficient should be in order to indicate acceptable reliability for a criterion-referenced test. Unfortunately, there is no single, simple answer to this question, as we learned in Chapter 4 during our discussion of the numbers of test items that should be included on a test. When we ask how high phi should be, we are asking, "How reliable does my test have to be?" Remember that perfect consistency would result in a phi coefficient of +1.00. The answer to this question depends entirely on what the consequences of testing error are (see Chapters 4 and 9). A test that assesses highly critical competencies that determine health and safety of clients or employees or that are essential to organizational survival should have phi reliability estimates above +.95. Tests of important competencies should have coefficients above +.75. Under no circumstances should tests with coefficients below +.50 be considered reliable tests.

PRACTICE

Table 11.2 shows example test data from two administrations of a criterion-referenced test. Practice using the formula presented above to calculate phi. A blank matrix (Figure 11.4) is provided to assist you.

**FIGURE 11.4 Blank Table for Practice Phi Calculation
Test-Retest Reliability**

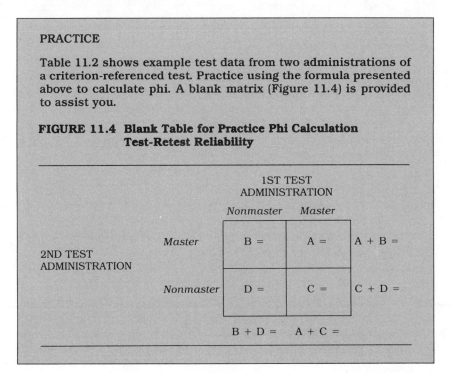

TABLE 11.2 Sample Test-Retest Data

1ST TEST ADMINISTRATION		2ND TEST ADMINISTRATION	
Test-Taker #	*Status*	*Test-Taker #*	*Status*
1	Master	1	Nonmaster
2	Master	1	Master
3	Master	3	Nonmaster
4	Master	4	Nonmaster
5	Master	5	Master
6	Master	6	Master
7	Nonmaster	7	Nonmaster
8	Nonmaster	8	Nonmaster
9	Nonmaster	9	Nonmaster
10	Nonmaster	10	Master
11	Nonmaster	11	Master
12	Master	12	Master

FEEDBACK

Your correctly completed matrix should look like the one in Figure 11.5.

FIGURE 11.5 Answer for Practice Phi Calculation Test-Retest Reliability

		1ST TEST ADMINISTRATION		
		Nonmaster	*Master*	
2ND TEST ADMINISTRATION	*Master*	B = 2	A = 4	A + B = 6
	Nonmaster	D = 3	C = 3	C + D = 6
		B + D = 5	A + C = 7	

Placing the numbers in the formula, you should have written:

$$\phi = \frac{[(4)(3)] - [(2)(3)]}{\sqrt{(6)(6)(7)(5)}}$$

$$= \frac{(12 - 6)}{\sqrt{1260}}$$

$$= \frac{6}{35.5}$$

$$\phi = .169 \text{ or } .17$$

The resulting phi coefficient is +.169 or +.17. This outcome indicates a very unreliable test.

The Agreement Coefficient

Description of the Agreement Coefficient. Not all test-retest reliability coefficients are correlation coefficients. One of the easiest reliability coefficients to understand is the agreement coefficient; the symbol for this statistic is p_o. Simply put, the agreement coefficient is the number of test-takers consistently classified on the two test administrations divided by the total number of test-takers. As such, the agreement coefficient is simply a percentage of consistent classifications achieved by the test.

Before we discuss the calculation of the agreement coefficient, it should be noted that it can result in deceptively high indices of reliability. A proportion of the consistent matches in the two test administrations would occur due to chance alone; these chance matches are included in the agreement coefficient. Remember that unlike phi, p_o is not a correlation coefficient and does not have the same range of values. A test that appears to have a decent agreement coefficient may have a very low phi coefficient. We will make such a comparison after we calculate p_o.

Calculating the Agreement Coefficient. The formula for calculating the agreement coefficient is

$$p_o = \frac{(a + d)}{N}$$

where **a** is the number of test-takers classifed as master on both administrations; **d** is the number of test-takers classified as non-master on both administrations; and **N** is the total number of test-takers.

Below is a calculation of the agreement coefficient for the test-retest data from Table 11.1 presented earlier.

$$p_o = \frac{(4 + 4)}{10}$$

$$= \frac{8}{10}$$

$$p_o = .80$$

PRACTICE

Use the p_o formula and the test-retest data from the practice phi calculation (Table 11.2) presented earlier to practice calculating the agreement coefficient.

FEEDBACK

The correct calculation of p_o for the test-retest data in Table 11.2 is as follows:

$$p_o = \frac{(4 + 3)}{12}$$

$$= \frac{7}{12}$$

$$p_o = .583 \text{ or } .58$$

How High Should the Agreement Coefficient Be?   Notice the sizable differences between the phi coefficients for the example data sets (the data in Tables 11.1 and 11.2) and their corresponding agreement coefficients. The phi coefficient for the first data set (Table 11.1) was .60 while the agreement coefficient was .80. The phi coefficient for the second data set (Table 11.2) was .17; the corresponding agreement coefficient is .58. It is essential for test developers who use these indices to realize that it would be inappropriate to compare the phi coefficient for one test with the agreement coefficient for another as if they were examining the two tests on a single reliability measure. The two statistics are quite different indices of reliability.

Of the two, we recommend phi because the agreement coefficient is inflated by matches in master/nonmaster status due simply to chance agreement alone. In other words, even a totally unreliable ⸍est will have a deceptively large number of matches due merely to ⸍nce. Therefore, the agreement coefficient can instill false confidence in the reliability of a test. The next coefficient presented, the kappa coefficient, was designed to overcome this weakness in the agreement coefficient. Because agreement coefficients tend to be inflated, it is recommended that for important master/nonmaster classification decisions, a test should achieve a p_o reliability coefficient above .85. For critical assessments, an agreement coefficient above .95 should be sought. It is sobering to remember that even with an agreement coefficient of .95, five test-takers out of a hundred are being inconsistently classified—conceivably nonmasters being certified as masters.

The Kappa Coefficient

Description of Kappa. The kappa coefficient (κ) is a refinement of the agreement coefficient in that it represents the test's improvement in master/nonmaster classification accuracy beyond the level of chance classification matches. The coefficient literally represents the proportion of possible improvement in classification accuracy achieved by the test.

The first task in figuring the kappa coefficient is to determine the agreement coefficient as already described. Subtracted from this agreement coefficient is the level of agreement expected due to chance. The result of this subtraction is then divided by 1 minus the agreement expected due to chance, which is the maximum possible improvement in accuracy that the test could make. Therefore, κ is a simple proportion; it is the actual improvement in classification consistency attributed to the test divided by the possible improvement in classification consistency.

Calculating the Kappa Coefficient. The formula for calculating the kappa coefficient is as follows:

$$\kappa = \frac{p_o - p_{chance}}{1 - p_{chance}}$$

The formula for p_o is, as presented earlier,

$$p_o = \frac{(\mathbf{a} + \mathbf{d})}{\mathbf{N}}$$

The formula for p_{chance} is

$$p_{chance} = \frac{[(\mathbf{a}+\mathbf{b})(\mathbf{a}+\mathbf{c})] + [(\mathbf{c}+\mathbf{d})(\mathbf{b}+\mathbf{d})]}{\mathbf{N}^2}$$

Remember that **N** equals the total number of test-takers. The following matrix (Figure 11.6)—similar to the one used to assist in the phi calculation—will demonstrate how to determine quickly and accurately the values for **a, b, c,** and **d** for calculating p_o and p_{chance}. (**NOTE:** This table is similar to the phi matrix, but the positions of the test administrations have been reversed and the master/non-master categories are located differently. Be careful to set up these matrices correctly for each of the two different reliability coefficients.)

FIGURE 11.6 Matrix for Determining p_o and p_{chance}

		2ND TEST ADMINISTRATION		
		Master	*Nonmaster*	
	Master	a =	b =	(a + b) =
1ST TEST ADMINISTRATION				
	Nonmaster	c =	d =	(c + d) =
		(a + c) =	(b + d) =	

Using the test-retest data originally presented in Table 11.1, an example of a kappa coefficient calculation follows. We begin by filling in the p_o and p_{chance} values table as illustrated in Figure 11.7 on page 181.

Next we figure p_o.

$$p_o = \frac{(4 + 4)}{10}$$

$$= \frac{8}{10}$$

$$p_o = .80$$

And p_{chance}.

$$p_{chance} = \frac{[(4+1)(4+1)] + [(1+4)(1+4)]}{10^2}$$

$$= \frac{[(5)(5)] + [(5)(5)]}{100}$$

$$= \frac{(25 + 25)}{100}$$

$$= \frac{50}{100}$$

$$p_{chance} = .50$$

FIGURE 11.7 Matrix for Determining p_o and p_{chance}

		2ND TEST ADMINISTRATION		
		Master	*Nonmaster*	
1ST TEST ADMINISTRATION	*Master*	a = 4	b = 1	(a + b) = 5
	Nonmaster	c = 1	d = 4	(c + d) = 5
		(a + c) = 5	(b + d) = 5	

Now that we have the values for p_o and p_{chance}, it is a simple matter to calculate the kappa coefficient.

$$\kappa = \frac{.80 - .50}{1 - .50}$$

$$= \frac{.30}{.50}$$

$$\kappa = .60$$

How High Should the Kappa Coefficient Be? It is important to remember that the coefficient of agreement and the kappa coefficient are two very different indices of reliability with distinct interpretations. The agreement coefficient is an indication of overall consistency in the test's classifications, whereas the kappa coefficient indicates the improvement in consistency over chance matches resulting from using the test. Therefore, one might expect that kappa values would be lower than coefficient of agreement values; however, when the test is perfectly reliable, both the agreement coefficient and the kappa coefficient will be equal to 1.00. (Kappa values and phi values tend to be similar, though kappa is not a correlation coefficent and so their interpretations are also distinct.)

Interpretations of how high a kappa coefficient should be differ among authorities. However, we tend to err on the side of higher standards than lower. Furthermore, we assume that the time and effort to establish test-retest reliability will not be expended unless the objectives assessed by the test are important to the organization. Therefore, we recommend that a kappa coefficient be above .60, and higher as the objectives become more critical.

PRACTICE

Using the test-retest data from Table 11.2 and the formulas for calculating the kappa coefficient above, practice calculating kappa for this data. A blank p_o and p_{chance} values matrix is provided to assist you (Figure 11.8).

FIGURE 11.8 Blank Matrix for Determining p_o and p_{chance}

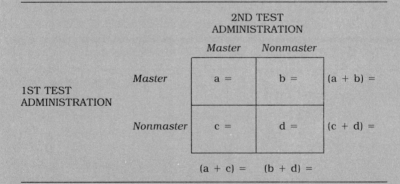

FEEDBACK

Your correctly completed p_o and p_{chance} values matrix should look like Figure 11.9 on page 183.

Your calculation of p_o should have been as follows:

$$p_o = \frac{(4 + 3)}{12}$$

$$= \frac{7}{12}$$

$$p_o = .58$$

The correct calculation of p_{chance} is

$$p_{chance} = \frac{[(4+3)(4+2) + (2+3)(3+3)]}{12^2}$$

$$= \frac{[(7)(6) + (5)(6)]}{144}$$

$$= \frac{(42 + 30)}{144}$$

$$= \frac{72}{144}$$

$$p_{chance} = .50$$

Having the values for p_o and p_{chance}, the kappa coefficient should have been calculated as follows:

$$\kappa = \frac{.58 - .50}{1 - .50}$$

$$= \frac{.08}{.50}$$

$$\kappa = .16$$

This outcome indicates an unreliable test (and is consistent with the phi coefficient determination of reliability calculated earlier).

FIGURE 11.9 Completed Practice Matrix for Determining p_o and p_{chance}

		2ND TEST ADMINISTRATION		
		Master	*Nonmaster*	
1ST TEST ADMINISTRATION	*Master*	a = 4	b = 3	(a + b) = 7
	Nonmaster	c = 2	d = 3	(c + d) = 5
		(a + c) = 6	(b + d) = 6	

Summary Comparison of ϕ, p_o, and κ

It is difficult to overstate the importance of understanding that these three reliability coefficients, while they all address test-retest reliability, are different indices requiring separate interpretations. However, if all three indices are calculated on the same test-retest data, the outcomes will, of course, be related. In other words, while a test with borderline reliability might clear the standard of acceptability for one of the indices and not for the others, tests will not look reliable on one index and unreliable on the next. Of the three, we recommend phi for the following reasons:

- While by far the easiest coefficient for most people to understand, the agreement coefficient, p_o, tends to be inflated by chance matches as discussed in some detail above. This characteristic makes it a dangerous coefficient in the hands of those who do not realize this weakness and encourages the naive implementation of unreliable tests. (It is a sobering exercise to calculate the agreement coefficient on test results from a test with a phi equal to 0 and a kappa coefficient equal to 0; in other words, on a totally unreliable test, you will find that a deceptively high number of matches appear due solely to chance.)

- While the kappa coefficient, κ, corrects this weakness in the agreement coefficient by reflecting only the improved accuracy attributable to the test, kappa thereby becomes, in our judgment, somewhat difficult to interpret intuitively. It just doesn't seem to make as much common sense as a correlation coefficient to most people.

It is worth noting again that all of the coefficents discussed can be expected to yield unstable results when calculated on small samples of test-takers. One of the most important pieces of advice we can give you in establishing the reliability (and validity) of your test is to use a sample size of at least 30 if at all possible.

TYPES OF VALIDITY

As mentioned above, reliability is a necessary but not sufficient condition for validity in testing. Establishing reliability assures consistency; establishing validity assures that the test consistently measures what it is supposed to measure. As is the case with reliability, there are several different types of validity. We shall present four types of validity:

- face validity
- content validity
- concurrent validity
- predictive validity

Each of the four types of validity are described below. Of these four, only the latter three are typically assessed formally. The descriptions of these validities—content, concurrent, and predictive—are followed by procedures for establishing each.

Face Validity

Description of Face Validity. The concept of face validity is best understood from the perspective of the test-taker. A test has face validity if it appears to test-takers to measure what it is supposed to measure. For the purposes of defining face validity, the test-takers are not assumed to be content experts. The legitimate purpose of face validity is to win this acceptance of the test among test-takers. This is not an unimportant consideration, especially among tests with significant and highly visible consequences for the test-taker. Test-takers who do not do well on tests that lack face validity may be more litigation prone than if the test appeared more valid.

In reality, criterion-referenced tests developed in accordance with the guidelines suggested in this book are not likely to lack face validity. If the objectives for the test are taken from the job or task analysis (Chapter 3), and if the test items are then written to maximize their fidelity with the objectives (Chapter 4), the test will almost surely have strong face validity. Norm-referenced tests that use test items selected primarily for their ability to separate test-takers rather than items grounded in competency statements are much more likely to have face validity problems.

It is important to note that while face validity is a desirable test quality, it is not adequate to establish the test's true ability to measure what it is intended to measure. The other three types of validity are more substantive for that purpose.

Content Validity

Description of Content Validity. A test possesses content validity when a group of recognized content experts or subject matter experts has verified that the test measures what it is supposed to measure. Note the distinction between face validity and content validity; content validity is formally determined and reflects the judgments of experts on the content or competencies assessed by the test, whereas face validity is an impression of the test held among non-experts.

Determining Content Validity. The steps in the process of determining content validity are described below.

1. The first step in establishing content validity is to select three to five judges who are experts in the competencies assessed by the test. If the test covers sufficiently unrelated objectives, you might have to have a panel of judges for subsets of the items. You might have to have more judges if the test covers sufficiently general objectives. For example, if the test were an assessment of management skills, you might have to have judges who could represent the major divisions of the organization—technical, operations, sales, etc.—to ensure that the test will be acceptable to managers throughout the organization. The identity of the judges and their credentials for serving as judges should be recorded for documentation purposes. This information could be important if the content validity of the test is ever challenged.

2. The judges are then presented with the objectives the test is supposed to assess and the items corresponding to each of these objectives. For each item the judges must decide whether or not the item assesses the intended objective. We recommend asking judges to make a yes/no decision regarding whether or not the item matches the objective rather than asking them to rate the objective on a scale. This recommendation simplifies the process for the judges, improves the reliability of their judgments, and facilitates the aggregation of the judges' opinions. Judges should also be asked if they see any technical problems with the item— any cuing of the correct answer, more than one possible correct answer, etc. Judges should also be provided with space to make any additional comments about the item that they think test developers ought to know.

3. It is suggested that judges review and rate the items independently first, then debrief their results together with the assis-

tance of one of the test's writers. The test writer should be there to hear firsthand the judges' remarks and concerns; this person can also facilitate the reaching of consensus among the judges regarding the acceptability of each item.

It should be noted that it is important that the objectives given to judges be based on an accurate job analysis. Since judges are only matching items to the objectives presented to them, they cannot be expected to discover a faulty job analysis. If the job analysis reveals more skills than the planned test can assess, it is important that the objectives chosen for inclusion be representative of the job and that the procedure used to select the objectives be documented in the event of legal challenges to the test's validity.

Figure 11.10 illustrates a sample form that you might use to collect the content validity judgments. Figure 11.11 illustrates how you might aggregate the opinions of the content validity judges.

Unlike reliability calculations and the other two validity procedures we will present, content validation does not result in a single numerical outcome that can be compared to a standard. As indicated above, if possible, it is advisable to have your judges reach consensus regarding the revisions required to make each item content valid. If the judges cannot be together to work as a group, it is recommended that you collect their individual opinions and examine any item that any judge felt did not match its objective and every item that was marked as technically flawed. You may, of course, have to call some of the judges back to get their approval on items that required substantial revision.

Concurrent Validity

Description of Concurrent Validity. The most desirable way to establish the validity of your test is to demonstrate its concurrent validity. Concurrent validity refers to the ability of a test to correctly classify masters and nonmasters. This is, of course, what you *hope* every criterion-referenced test will do; however, certainly face validation and even content validation do not actually demonstrate the test's ability to classify correctly. Concurrent validation is the technical process that allows you to evaluate the test's ability to distinguish between masters and nonmasters of the assessed competencies.

Determining Concurrent Validity. The process of calculating a test's concurrent validity is similar in some ways to the process

FIGURE 11.10 Test Content Validation Form

Judge: _____   Course: _____

Title: _____   Course Number: _____

Location: _____   Test Number: _____

_____   Date: _____

Please read each objective and its corresponding items. For each test item, please make two judgments.

1. Do you feel that the item assesses its intended objective? Circle "Y" for "Yes" or "N" for "No" to indicate your opinion. If you are uncertain, circle "N" and explain your concern in the comments section.

2. Do you see any technical problem with the item? For example, is there more than one correct answer among the alternatives? Is there a cue to the correct answer within the item? Is the indicated correct answer indeed correct? etc. Circle "OK" if you see no problems; circle the "?" if you do see technical problems and explain your concern in the comments section.

Please feel free to add any additional comments you think would be helpful to the designers of this test.

Item Opinion Record

Objective #	Item #(s)	Match	Technical Problems		Comments
1	3	Y N	OK	?	
	8	Y N	OK	?	
	10	Y N	OK	?	
2	1	Y N	OK	?	
etc.					

FIGURE 11.11 Summary of Content Validity Judgments

Objective #	Item #	Judge 1 2 3	% of Matches	Technical Problems/ Comments
1	3	Y Y Y	100	None noted
	8	Y Y N	66	Stem is unrealistic.
	10	Y Y Y	100	Cue in distractor "b"

used to assess its test-retest reliability. However, only one test administration is required to calculate concurrent validity. The success of the procedure depends heavily on your ability to form the group of sample test-takers correctly. The steps in the process are described below.

1. The first step is to identify a group of masters and a group of nonmasters to serve as your validation sample test-takers. It is *not sufficient* to form a group that you think in all likelihood contains some masters and some nonmasters. In order to establish concurrent validity, you must know before the test is administered the master/nonmaster status of every individual test-taker in the group. Furthermore, the sample test-takers should be representative of the masters and nonmasters between whom you want the test to distinguish when it is finally implemented by the organization. Errors here are easy to make. For example, including nonmasters in your validation sample who are far less competent than those nonmasters who will ultimately take the test will not tell you whether your test will classify correctly when actually implemented. By including less competent nonmasters in your validation sample, you have in effect made the classification task too easy for your test during the validation—easier than the classification decisions the test must make when actually used by the organization to distinguish between masters and nonmasters.

    It is also important that the validation sample be representative of the demographic characteristics (race, sex, national origin, etc.) of the future test-takers. If the test's concurrent validity were legally challenged, it would be important to document that it is a valid instrument for all groups; this could be difficult to do if the test has been validated only on a sample of white males, for example.

    This group of sample test-takers should be sizable—at least forty, more if you can get them—about equally composed of masters and nonmasters. For documentation purposes you should keep a record of who the test-takers were and how they were selected; this information could be important if the concurrent validity of the test were challenged at a later date.

2. The next step is to administer the test to this group and record how the test classifies each test-taker. The test should be administered exactly as it will be when implemented.

3. A phi, ϕ, correlation coefficient is then calculated on the two sets of master/nonmaster classifications, i.e., the known status

and the status according to the test for each member of the
sample.

As you will recall from the earlier section on reliability, phi is a
correlation coefficient used to determine the relationship between
two dichotomous variables—two variables that have only two values
each. In the case of concurrent validity the two variables are the test-
taker's known mastery status and the test-taker's mastery status
according to the test. The two values on each of these variables are
"master" and "nonmaster." Like all correlation coefficients, the size
of phi can range from +1.00 to −1.00. See the earlier section in
this chapter on correlation for an explanation of the meaning of
direction and magnitude of correlation coefficients. Table 11.3 contains an example of the results of concurrent validation procedures.

In order to calculate the phi coefficient, the data should first be
placed in a phi matrix table set up like the one in Figure 11.12. The
values in the phi matrix for concurrent validity are: **B** is the number
of test-takers who are known to be masters, but whom the test
classified as nonmasters; **A** is the number who are known to be
masters and whom the test identified as masters; **D** is the number
known to be nonmasters whom the test identified as nonmasters;
and **C** is the number known to be nonmasters, but whom the test
classified as masters.

TABLE 11.3 Example of Concurrent Validity Data

Test-Taker #	Known Status	Test-Taker #	Test Status
1	Master	1	Master
2	Master	2	Master
3	Master	3	Master
4	Master	4	Master
5	Master	5	Master
6	Master	6	Master
7	Nonmaster	7	Nonmaster
8	Nonmaster	8	Nonmaster
9	Nonmaster	9	Nonmaster
10	Nonmaster	10	Nonmaster
11	Nonmaster	11	Nonmaster
12	Nonmaster	12	Master

FIGURE 11.12 Phi Table for Concurrent Validity

		TEST STATUS		
		Nonmaster	*Master*	
KNOWN STATUS	*Master*	B =	A =	A + B =
	Nonmaster	D =	C =	C + D =
		B + D =	A + C =	

The formula for calculating phi using the values given is

$$\phi = \frac{(AD) - (BC)}{\sqrt{(A + B)\,(C + D)\,(A + C)\,(B + D)}}$$

Figure 11.13 shows how the matrix (Figure 11.12) would appear correctly completed with the sample data from Table 11.3. Substituting the numbers from the matrix into the phi formula, we can complete the calculation of the test's concurrent validity coefficient.

$$\phi = \frac{[(6)(5)] - [(0)(1)]}{\sqrt{(6)(6)(7)(5)}}$$

$$= \frac{(30 - 0)}{\sqrt{1260}}$$

$$= \frac{30}{35.5}$$

$$\phi = .845 \text{ or } .85$$

FIGURE 11.13 Example Phi Table for Concurrent Validity

		TEST STATUS		
		Nonmaster	*Master*	
KNOWN STATUS	*Master*	B = 0	A = 6	A + B = 6
	Nonmaster	D = 5	C = 1	C + D = 6
		B + D = 5	A + C = 7	

The concurrent validity of this example test is .85. This outcome indicates a test with fairly high validity. By inspecting the sample data, you can see that only one test-taker was misclassified by the test. If all test-takers had been correctly classified, the phi coefficient would have been 1.00, indicating a perfect correlation between the known status of the test-takers and their classification by the test. The reason that a single misclassified test-taker lowers the validity coefficient so much is due to the small sample size. A single misclassification within a sample of 40 test-takers would result in a far higher validity coefficient—.95, to be exact.

As in deciding how high a reliability coefficent should be, the required magnitude of the validity coefficent depends on the criticality of the assessment, i.e., on what the consequences to the organization are of making classification errors. For critical objectives you would want your concurrent validity coefficient to be .95 or above. For important objectives, it should be above .75. Tests with coefficients below .50 should not be considered valid for any purpose.

PRACTICE

TABLE 11.4 Sample Concurrent Validity Data

Test-Taker #	Known Status	Test-Taker #	Test Status
1	Master	1	Master
2	Master	2	Master
3	Master	3	Master
4	Master	4	Nonmaster
5	Master	5	Nonmaster
6	Master	6	Nonmaster
7	Nonmaster	7	Nonmaster
8	Nonmaster	8	Nonmaster
9	Nonmaster	9	Nonmaster
10	Nonmaster	10	Nonmaster
11	Nonmaster	11	Nonmaster
12	Nonmaster	12	Master

Table 11.4 contains practice data resulting from concurrent validation procedures. Using these data and the formula for phi above,

practice calculating the concurrent validity coefficient for these example test data. A blank phi matrix (Figure 11.14) is provided to assist you.

FIGURE 11.14 Blank Table for Practice Phi Calculation Concurrent Validity

		TEST STATUS		
		Nonmaster	*Master*	
KNOWN STATUS	*Master*	B =	A =	A + B =
	Nonmaster	D =	C =	C + D =
		B + D =	A + C =	

FEEDBACK

Your correctly completed matrix should look like the one in Figure 11.15.

FIGURE 11.15 Answer for Practice Phi Calculation Concurrent Validity

		TEST STATUS		
		Nonmaster	*Master*	
KNOWN STATUS	*Master*	B = 3	A = 3	A + B = 6
	Nonmaster	D = 5	C = 1	C + D = 6
		B + D = 8	A + C = 4	

Placing the numbers in the formula, you should have written:

$$\phi = \frac{[(3)(5)] - [(3)(1)]}{\sqrt{(6)(6)(4)(8)}}$$

$$= \frac{(15 - 3)}{\sqrt{1152}}$$

$$= \frac{12}{33.9}$$

$$\phi = .353 \text{ or } .35$$

The result is that the concurrent validity coefficient for this practice test data is .35, indicating an invalid test.

Predictive Validity

Description of Predictive Validity.   Another type of validity frequently confused with concurrent validity is predictive validity. There is an important conceptual distinction between the two and the procedures for calculating them. Whereas concurrent validity means that a test can correctly classify test-takers of currently known competence, predictive validity means that a test can accurately predict future competence. Predictive validity is important for many personnel selection devices that are used to choose persons for specific job responsibilities. Tests used to help persons select careers also require high predictive validity. In both of these cases the test is taken first, while the demonstration of competence—job performance or successful career achievement—comes later; hence the term predictive validity.

Determining Predictive Validity.   Predictive validity is determined in much the same way as concurrent validity. However, as indicated in the description above, the pool of test-takers must first take the test. At some later point in time, after their competence on the job can be determined, their achieved mastery status is correlated with their earlier performance on the test. A phi correlation coefficient could be used to calculate predictive validity in the same way it was used to calculate concurrent validity above. The statistic would correlate test status—master or nonmaster—with future achieved status—master or nonmaster. Figure 11.16 illustrates the phi matrix for determining predictive validity. The formula for calculating phi is the same as above.

FIGURE 11.16 Phi Table for Predictive Validity

		TEST STATUS		
		Nonmaster	*Master*	
FUTURE STATUS	*Master*	B =	A =	A + B =
	Nonmaster	D =	C =	C + D =
		B + D =	A + C =	

SUMMARY COMMENT ON RELIABILITY AND VALIDITY

It should be apparent that it requires an investment of time and resources to establish the reliability and validity of a test. It should be noted that one way to streamline the process is to form the sample test-takers for the concurrent validation process—the group of known masters and nonmasters—and arrange to have them take the test twice, in order to establish the test's reliability in making correct classifications over time, i.e., test-retest reliability.

It should also be noted, however, that it is difficult to overstate the importance of reliability and validity for tests of any kind. Tests that are not reliable and valid are, by definition, not worth giving. Many organizations are willing to assume that their tests are reliable and valid. These organizations are in jeopardy from two sources:

- Unreliable, invalid tests cannot be expected to be upheld in a court of law should the fairness of the tests be challenged.
- Even if the tests are never legally challenged, the costs of bad decisions resulting from unreliable, invalid tests can be extremely high; not only are such tests useless because they do not provide the information regarding competence that they are presumed to provide, but they actively encourage the organization to misplace people, resources, time, and talent.

PART FIVE

LEGAL ISSUES IN CRITERION-REFERENCED TESTING

12.

A Practitioner's Guide to Criterion-Referenced Test Development and the Law

A SHORT HISTORY OF THE *UNIFORM GUIDELINES ON EMPLOYEE SELECTION PROCEDURES*
SUMMARY AND INTERPRETATION OF THE EEOC
QUESTIONS AND ANSWERS ABOUT THE *GUIDELINES*

A SHORT HISTORY OF THE *UNIFORM GUIDELINES ON EMPLOYEE SELECTION PROCEDURES*

In reality, there is no unified code of laws regarding testing. However, there are a set of guidelines that have come to be viewed with "great deference" by the courts and have come to have the effect of law. These guidelines are the *Uniform Guidelines on Employee Selection Procedures (Guidelines)* adopted by the Equal Opportunity Commission in 1978 (U.S. Equal Opportunity Commission et al.). The *Guidelines* were an outgrowth of Title VII of the 1964 Equal Employment Opportunity Act and were designed to protect individuals from employment discrimination.

The *Guidelines* provides definitions of discrimination and adverse impact, information on how adverse impact is determined, standards for conducting validity studies, and related validation issues such as the use of alternative selection methods, cooperative validation studies, fairness evidence, and the kinds of documentation of adverse impact and validity evidence the user needs to collect and maintain. (Nathan and Cascio, in Berk, 1986, p. 12)

The *Guidelines* were quickly adopted by federal agencies and thus became the *de facto* standard for evaluating the testing decisions that led to discriminatory hiring patterns. Shortly thereafter, in 1979, a set of 90 questions and answers were issued to clarify common questions about the meaning of the *Guidelines.* An additional three questions and answers were added in 1980 (U.S. Equal Opportunity Commission et al.). These questions did not alter the meaning of the *Guidelines* but were purely an attempt to make them more understandable.

Theoretically, the *Guidelines* are concerned only with issues of discrimination in hiring. In reality, they have served as the focal point for many court decisions on testing, criterion-referenced and otherwise, for example, performance appraisals, assessment centers, etc. Because of the legal precedents they have created, an understanding of the *Guidelines* is the single most useful legal perspective for a test designer. While there has been some controversy about the current usefulness of the *Guidelines* (Ballew, 1987; Smith, 1985a) there really is no other single, workable alternative to guide test development. Thus our approach in this chapter will be to summarize the content of the 1979 and 1980 questions and answers that were adopted to provide a common interpretation of the *Guidelines* and to integrate our own comments. The headings we will use are the headings used by the EEOC in the questions and answers. The summary and comments represent our professional judgment and should not be taken as legal advice. If you have a specific concern, it is best to consult with your organization's legal counsel.

SUMMARY AND INTERPRETATION OF THE EEOC QUESTIONS AND ANSWERS ABOUT THE *GUIDELINES*

I. Purpose and Scope

The *Guidelines* are designed to promote the goal of equal opportunity of employment regardless of "race, color, sex, religion, or national origin" (p. 11997). They were designed to provide a uniform set of standards for the development of any form of test that would affect employment opportunities, i.e., standards that are consistent with generally accepted practices of psychological test construction. The *Guidelines* apply to nearly any organization and cover employee selection for

> hiring, retention, promotion, transfer, demotion, dismissal or referral (and apply to the use of) job requirements (physical, educa-

tion, experience), and evaluation of applicants or candidates on the basis of interviews, performance tests, paper-and-pencil tests, performance in training programs or probationary periods, and any other procedures used to make an employment decision whether administered by the employer or by an employment agency. (p. 11997)

The *Guidelines* are designed to encourage the fair treatment of protected groups in our society, but they are not sanction for incompetence. To disproportionately screen out a group is illegal, unless

> the process or its component procedures have been validated in accord with the *Guidelines*, or the user otherwise justifies them in accord with Federal law. (p. 11997)

In other words, a reliable and valid test will stand up in court regardless of the outcome for protected groups. The systematic design of tests, as we have described it, should lead not only to a legally defensible position, but to professional and ethical behavior.

II. Adverse Impact, the Bottom Line, and Affirmative Action

The *Guidelines* do not require that validated tests be used in all situations. The authors note, as we have, that such procedures are desirable, but that evidence of validity will need to be provided only when the selection procedure adversely affects those of a given race, sex, or ethnic group. However, mounting case law would appear to support the need for evidence of reliable and valid tests should a hiring decision based on a test be challenged.

Adverse Impact. Adverse impact is a specific term in the *Guidelines* and is defined by the following two criteria:

1. A selection rate of less than 80% for a given race, sex, or ethnic group in comparison to the highest selection rate in the pool.
2. When a given group is more than 2% of the labor force in the labor area.

For example, if the local labor force was 70% White and 30% Black, you would need to maintain records for these two groups. If there were 120 applicants for a job, 80 White and 40 Black, equally divided by sex; and you hired 48 Whites (60%) and 12 Blacks (30%), you

TABLE 12.1 Sample Summary of Adverse Impact Figures

			Black	White	Total
			Race		
			Black	*White*	*Total*
Sex	Male	Pass	6	24	30
		Fail	14	16	30
	Female	Pass	6	24	30
		Fail	14	16	30
Total		Pass	12	48	60
		Fail	28	32	60

would not be in compliance with the *Guidelines* because the Black hire rate at 30% is less than 48% (48% being 80% of the highest selection rate of 60%—for Whites). Table 12.1 summarizes this situation.

In this example, the Black selection rate is 30% (12/40) while the White selection rate is 60% (48/80). Under the definition of adverse impact, any selection rate that is less than 80% of the highest rate is discriminatory. Since the highest selection rate for races is for Whites at 48/80 or 60%; then 80% of 60% = 48%. Therefore, any hiring rate less than 48% will be viewed as racially discriminatory. In this instance, the hiring rate for Blacks is 30% and would thus be in violation of the *Guidelines*. There is no difference in the hiring rate by sex, both rates being 30/60, or 50%, so no adverse impact would be indicated for these groups.

PRACTICE

Review Table 12.2 and determine if there is evidence of adverse impact by race or sex.

TABLE 12.2 Practice Adverse Impact Figures

		Black	White	Hispanic	Total
				Race	
Sex	Male Pass	20	5	5	30
	Male Fail	30	10	10	50
	Female Pass	10	10	0	20
	Female Fail	20	10	10	40
Total	Pass	30	15	5	50
	Fail	50	20	20	90

FEEDBACK

There is evidence of adverse impact on Hispanics. The evidence was determined as follows:

1. Determine the passing ratios for each protected group.
   Passing ratio for Black = 30/80 or 38%
   Passing ratio for Whites = 15/35 or 43%
   Passing ratio for Hispanics = 5/25 or 20%
   Passing ratio for Males = 30/80 or 38%
   Passing ratio for Females = 20/60 or 33%
   The highest passing ratio by race is 43%; by sex, 38%.

2. Determine the minimal selection rate for each protected group. To do this for race, you multiply the highest selection rate (in this case, that for Whites at 43%) by 80% or .80.
   a. Selection rate by race = 80% of 43%.
   $$= .80(.43)$$
   $$= .34 \text{ or } 34\%$$
   At 25% the Hispanic selection rate is below the minimal selection rate of 34%; therefore adverse impact is indicated. At 38% the Black selection rate is above the minimal selection rate required and, therefore, would not be seen as evidence of adverse impact.
   To determine the minimal selection rate for sex, you multiply the highest selection rate (in this case, that for males at 38%) by 80% or .80.

b. **Selection rate by sex** = 80% of 38%
                              = .80(.38)
                              = .30 or 30%
The selection rate for females at 33% is above the minimal selection rate of 30% and would not be seen as evidence of adverse impact.

The Bottom Line. The "bottom line" is a specific term in the *Guidelines* that refers to the principle that federal agencies will not seek enforcement actions against an organization if the total selection procedure does not have adverse impact, i.e., a selection differential exceeding 80% of the highest protected group's rate. The 80% level is a rule of thumb and:

> is not intended as a legal definition, but is a practical means of keeping the attention of the enforcement agencies on serious discrepancies in rates of hiring, promotion, and other selection decisions. (p. 11998)

However, this does not necessarily mean that the organization will not be prosecuted, even if the sum total of all tests the organization uses are in compliance with the 80% rule. In *Connecticut v. Teal* (cited in Smith, 1985a, p. 23) the Supreme Court found that while a particular company's total bottom line was acceptable, the results of a specific test that had an adverse impact on Blacks was discriminatory. Nor is the bottom line limited only to the 80% rule, as the courts have often required evidence of statistical significance in addition to the differential selection rates among protected groups.

Adverse impact, the 80% rule, is best measured against the "total selection process" for a given job, i.e., all steps leading to the final choice of an applicant for a given job. If there are a series of tests—medical, psychological, etc.—used to make the final hiring decision, only the final outcome for the hiring decision need be determined. Adverse impact does not have to be calculated for each step along the way. Nor does adverse impact have to be determined if the number of selections is too small to warrant it. (The definition of "too small" is not provided in the *Guidelines*, although they do note that the regulations apply to most organizations with more than 15 employees for more than 20 weeks a year.) However, records on employment decisions must be kept on

> groups for which there is extensive evidence of continuing discriminatory practices . . . For groups for which records are not re-

quired, the person(s) complaining may obtain information from the employer or others (voluntarily or through legal process) to show that adverse impact has taken place. (p. 12000)

The specific collection of adverse impact data is a matter of organizational policy that should be developed by the legal department. There are a number of tough questions that you as an instructional designer or test developer are probably not in a position to resolve, but can at least raise with the appropriate division. For example,

- Are data to be collected for each course? Promotion? New hires?
- Who are the protected groups for whom the data are to be collected? Blacks, Whites, Hispanics, American Indians, Asian/Pacific Islanders?
- What are the boundaries of the labor force?
- What is the proportion of these groups in the labor force?
- While an annual summary of hiring decisions is usually required by the EEOC, should they choose to explore an organization's procedures for employee selection, how often should the data be compiled, e.g., quarterly, semi-annually, yearly?
- Who will be responsible for collecting the data and reporting it?

Affirmative Action. The *Guidelines* are reasonably straightforward about their relationship to affirmative action policies. They

encourage the development and effective implementation of affirmative action plans or programs in two ways. First, in determining whether to institute action against a user on the basis of a selection procedure which has adverse impact and which has not been validated . . . (and) Second . . . do not preclude the use of selection procedures, consistent with Federal law, which assist in the achievement of affirmative action objectives. (p. 12001)

The *Guidelines* (Section 6A) allow the use of "lawful" alternative selection procedures that eliminate adverse impact when the employer has not demonstrated the validity of the initial testing process; but they do "not impose a duty to adopt a hiring procedure that maximizes hiring of minority employees" (p. 12004). The search for alternatives is required only during the "validity study" (Smith, 1985a) for the test, i.e., during the job analysis, planning, and test validation stages. Smith (1985a) notes that:

the search for alternatives (testing strategies) is required only during the course of a validity study, and there need be only a reasonable investigation . . . reasonable, in most circumstances, (is) a search of the published literature. Investigation of the unpublished literature is required only when adverse impact is high and validity is low . . . The legal standard of *Moody*, under which a plaintiff can prove discrimination by showing that an alternative serving the employer's legitimate business needs but with lesser adverse impact exists, is not changed by the guidelines of the questions and answers. (p. 23)

Section II of the *Guidelines*, however, concludes with the following:

Nothing in Section 6A should be interpreted as discouraging the use of properly validated selection procedures; but Federal equal employment opportunity law does not require validity studies to be conducted unless there is adverse impact. (p. 12001)

We suppose there may be a temptation by some to review this final passage and conclude that the easiest way out is to set the cut-off score low enough that nearly all applicants pass the test. However, Cascio, Alexander, and Barrett (1988) observed:

The courts have recognized this to be an exercise in futility; if an organization is unable to weed out those individuals who are min-imally qualified, what possible justification is there for the use of a test in the first place? In fact, one could argue that selectees should be drawn at random from those who score above the cut-off score . . . Setting a very low cut-off score (one that might be called "ar-bitrarily low," or as one court said, "a remarkably humble level," since a person could pass the test by responding randomly) tends, in the opinion of the court, to destroy the credibility of the entire testing process. (pp. 6–7)

III. General Questions Concerning Validity and the Use of Selection Procedures

The *Guidelines* address three types of validity: criterion-related, content, and construct. Criterion-related validity (or concurrent va-lidity) shows a relationship between scores on a test and job perfor-mance from a sample of workers. Content validity demonstrates a logical link between test items and job skills. Construct validity es-tablishes a relationship between some underlying trait, character-istic, or ability, e.g., mechanical aptitude, and the successful job performance. Of these three types, content validity (Does the test match the job analysis?) and criterion-related validity (Does the test sort known masters and nonmasters?) are the primary concerns of trainers; construct validity studies are best left to psychometricians due to their reliance on more subtle statistical assumptions and analysis.

The job analysis for a test should be carried out before you plan the test, as we noted in Chapter 3. Furthermore, the validation of the test should be conducted before the test is implemented. According to the *Guidelines*, if you determine that a test without evidence of validity does have an adverse impact, and you continue to use the test

> until the procedure is challenged (users) increase the risk that they will be found to be engaged in discriminatory practices and will be liable for back pay, awards, plaintiff's attorney's fees, loss of Federal contracts, subcontracts or grants, and the like. Validation studies begun on the eve of litigation have seldom been found to be adequate. (p. 12002)

In conducting a validity study, you should also attend to a number of other issues.

- A validity study must not only be job specific, it may also need to be location specific. In other words, the *Guidelines* do not assume that a study conducted in one location, for example, New York City, will be a fair test for workers in another location, for example, Carbondale, Illinois. To be safe, the validity study should be conducted as a joint venture with other members of the organization who will be affected by the results of the test. Paraphrasing the *Guidelines*,

  Results from one site may be used elsewhere: 1) if the original study has been shown to be valid, 2) the job(s) are closely matched, 3) there is evidence of fairness in the original study, 4) any variables that might affect fair use in the new location have been considered. (p. 12006)

- The same method for determining validity does not have to be used for all parts of a test.

  For example, where a selection process includes both a physical performance test and an interview, the physical test might be supported on the basis of content validity, and the interview on the basis of a criterion-related study. (p. 12003)

- Users do not have to develop the test themselves. If another test will meet all standards, then it is appropriate to use that test. However, in the event of a challenge, proof will rest with the organization using the test even if a test manual claims validity for the test.

- Establishing the validity of the test is a necessary but not sufficient step in the selection process. The test in use must be the test as validated.

  For example, if a research study shows only that, at a given passing score the test satisfactorily screens out probable failures, the study would not justify the use of substantially different passing scores, or of a ranked list of those who passed. (p. 12003)

IV. Technical Standards

The choice of a validation technique can be made based on its appropriateness for the type of selection and its technical and administrative feasibility. Criterion-related (concurrent) validity studies will require an appropriate sample of adequate size of masters and nonmasters and the use of reliable and valid measures of job performance. Content validity is used when it is possible to develop measures of visibly related to job skills. The content for the test should be drawn only from knowledge, skills, or abilities associated with the job and should not test for behaviors that an employee "will be expected to learn on the job." (p. 12004)

  The phrase "on the job" is intended to apply to training which occurs after hiring, promotion or transfer. However, if an ability, such as a language, takes a substantial length of time to learn, is required for successful job performance, and is not taught in advance, a test for that ability may be supported on a content validity basis. (p. 12007)

Therefore, while the test should not cover skills to be learned later after the job has been taken, in some circumstances the test can appropriately assess abilities that a candidate must bring to the job.

Ranking of performers on a test is acceptable only if there is a demonstrated statistical relationship between the rank and levels of performance, i.e., ranking must rest on an inference that higher scores on the procedure are related to better job performance.

- For example, for a particular warehouse worker job, the job analysis may show that lifting a 50-pound object is essential, but the job analysis does not show that lifting heavier objects is essential or would result in significantly better job performance. In this case a test of ability to lift 50 pounds could be justified on a content validity basis for a pass/fail determination. However, ranking of candidates based on relative amount of weight that can be lifted would be inappropriate. . . .

- On the other hand, in the case of a person to be hired for a typing pool, the job analysis may show that the job consists almost entirely of typing from manuscript, and that productivity can be measured directly in terms of finished copy. For such a job, typing constitutes not only a critical behavior, but it constitutes most of the job. A higher score on a test which measured words per minute typed, with adjustments for errors, would therefore be likely to predict better job performance than a significantly lower score. Ranking or grouping based on such a typing test would therefore be appropriate under the *Guidelines*. (p. 12005)

V. Records and Documentation

An organization is required to keep records that detail testing impact by race, sex, or ethnic groups. (Again, the legal or personnel departments should be able to specify which groups are to be monitored.) The *Guidelines* do not specify a particular procedure or process for collecting and recording the data, so once again, your organization should have established regulations for the record-keeping process. While the *Guidelines* do make a distinction in record-keeping needs for small employers (under 100 employees) and others, any organization should be able to provide data on:

- the number of persons hired, promoted, and terminated for each job by sex, and where appropriate by race and national origin;
- the number of applicants for hire and promotion by sex, where appropriate by race and national origin; and
- the selection procedures utilized (either standardized or not standardized).

Larger employers should also be able to provide:

- Records on the total selection process for a job where there is evidence of adverse impact. These total records need not be kept if there is no evidence of adverse impact.
- A justification of any procedure that is used to determine adverse impact using procedures other than the 80% criterion ("four-fifths rule").
- When adverse impact has been eliminated in the total selection process for a job, records must be kept for that year and at least two years after.

- If there is insufficient data to determine impact due to a small number of selections during the year, records should continue to be kept until there is sufficient evidence to determine whether or not adverse impact has been demonstrated. (pp. 38303–38304)

Any validity studies used to develop the testing process will need to be reported. While the *Guidelines* do describe formats for reporting on the various types of validity studies, for example, content, concurrent, etc. (see Section 15 of the *Guidelines*), they all contain similar elements that have probably been reinterpreted by your organization. These elements include the following:

1. When and where the study was conducted.
2. A description of the selection procedure, how it is used, and the results by race, sex, and ethnic group.
3. How the job was analyzed or reviewed and what information was obtained from this job analysis or review.
4. The evidence demonstrating that the selection procedure is related to the job. The nature of this evidence varies, depending upon the strategy used:
   1. Criterion-Related Validity (Concurrent Validity)
      - A description of the criterion measures of job performance, how and why they were selected, and how they were used to evaluate employees.
      - A description of the sample used in the study, how it was selected, and the size of each race, sex, or ethnic group in it.
      - A description of the statistical methods used to determine whether scores on the selection procedure (master/non-master) are related to scores on the criterion measures of job performance, and the results of these statistical calculations.
   2. Content Validity
      - The content of the job, as identified from the job analysis.
      - A description of the procedure by which content was selected for inclusion on the test.
      - The evidence demonstrating that the tested content resulting from the selection procedure is a representative sample of the content of the job.
5. What alternative selection procedures and alternative methods of using the selection procedure were studied and the results of the study.

6. The name, address, and telephone number of a contact person who can provide further information about the study.

In summary, the *Guidelines* support a systematic approach to test development. They provide a framework for the creation of valid and reliable assessments of performance and to support the goals of affirmative action. However, they do not require an organization to establish quotas or accept unqualified applicants or candidates for a job. They are an attempt to operationalize the procedures of systematic test design and development—from the initial job analysis to the final determination of mastery/nonmastery. They are simply guidelines for professional behavior in test construction.

RESOURCES

(A reference list and selected bibliography)

AERA/APA/NCME Joint Committee. (1985). *Standards for educational and psychological testing.* Washington, DC: American Psychological Association.

Ballew, P.J. (1987). Courts, psychologists, and the EEOC's Uniform Guidelines: An analysis of recent trends affecting testing as a means of employee selection. *Emory Law Journal, 36,* 203–252.

Barrett, R.S. (1981). Is the test content-valid: Or, who killed Cock Robin? *Employee Relations Law Journal, 6(4),* 584–600.

Berk, R. A. (Ed.). (1980). *Criterion-referenced measurement: The state of the art.* Baltimore, MD: Johns Hopkins University Press.

Berk, R. A. (Ed.). (1984). *A guide to criterion-referenced test construction.* Baltimore, MD: Johns Hopkins University Press.

Berk, R. A. (Ed.). (1986). *Performance assessment methods and applications.* Baltimore, MD: Johns Hopkins University Press.

Bloom, B.S. (Ed.). (1956). *Taxonomy of educational objectives. Handbook I: Cognitive Domain.* New York: David McKay.

Cascio, W.F., Alexander, R.A., & Barrett, G.V. (1988). Setting cutoff scores: Legal, psychometric, and professional issues and guidelines. *Personnel Psychology, 41,* 1–24.

Conger, A.J. (1980). Integration and generalization of kappas for multiple raters. *Psychological Bulletin, 88(2),* 322–328.

Davis, R.H., Alexander, L.T., & Yelon, S.L. (1974). *Learning systems design.* New York: McGraw-Hill.

Dick, W., & Hagerty, N. (1971). *Topics in measurement: Reliability and validity.* New York: McGraw-Hill.

Edwards, A. (1973). *Statistical methods.* (3rd ed.). New York: Holt, Rinehart and Winston.

Gagné, R.M. (1985). *The conditions of learning.* (4th ed.). New York: Holt, Rinehart and Winston.

Gallagher, D.G., & Veglahn, P.A. (1986, October). Arbitral standards in cases involving testing issues. *Labor Law Journal, 37,* 719–730.

Guilford, J.P., & Fruchter, B. (1978). *Fundamental statistics in psychology and education* (6th ed.). New York: McGraw-Hill.

Haertel, E. (1985). Construct validity and criterion-referenced testing. *Review of Educational Research, 55(1),* 23–46.

Haney, C. (1982). Employment tests and employment discrimination: A dissenting psychological opinion. *Industrial Relations Law Journal, 5(1),* 1–86.

Keller, W.L. (1981, January). Defending before the EEOC. *For the Defense,* 10–23.

Kleiman, L.S., & Faley, R.H. (1985). The implications of professional and legal guidelines for court decisions involving criterion-related validity: A review and analysis. *Personnel Psychology, 38,* 803–833.

Kleiman, L.S., & Raley, R.H. (1978). Assessing content validity: Standards set by the court. *Personnel Psychology, 31,* 701–713.

Lathrop, R.J. (1983). The number of performance assessments necessary to determine competence. *Journal of Instructional Development, 6(3),* 26–31.

Lathrop, R.J. (1986). Practical strategies for dealing with unreliability in competency assessments. *Journal of Educational Research, 70 (4),* 234–237.

Livingston, S.A., & Zieky, M.J. (1982). *Passing scores.* Princeton, NJ: Educational Testing Service.

Mager, R.F. (1962). *Preparing instructional objectives.* Belmont, CA: Fearon Publishers.

Mills, C.N., & Melican, G.J. (1988). Estimating and adjusting cutoff scores: Features of selected methods. *Applied Measurement in Education, 1(3),* 261–275.

Merrill, M.D. (1983). Component Display Theory. In C.M. Reigeluth (Ed.). *Instructional design theories 2nd models: An overview of their current status.* Hillsdale, NJ: Lawrence Erlbaum Associates.

Nathan, B.R., & Cascio, W.F. (1986). Introduction. Technical and legal standards. In R.A. Berk (Ed.), *Performance assessment methods and applications.* Baltimore, MD: Johns Hopkins University Press, pp. 1–50.

National Computing Systems. (1988). *Microtest and examSystem.* Minneapolis, MN: National Computer Systems. (800-367-6627)

Noonan, J.V., & Sarvela, P.D. (1988). Implementation decisions in computer-based testing programs. *Performance & Instruction, 27(6),* 5–13.

Okey, J.R. (1973). Developing and validating learning hierarchies. *Audio-Visual Communications Review, 21(1),* 87–108.

Russell, J.S. (1984). A review of fair employment cases in the field of training. *Personnel Psychology, 37,* 261–276.

Scantron Corporation. (1988). *Optical mark reader/data terminals.* Tustin, CA: Scantron Corporation. (800-421-5066)

Smith, Jr., C. (1985a). The EEOC's standards for employment testing. *The National Law Journal,* September 16, pp. 22–23.

Smith, Jr., C. (1985b). Testing must relate to specific job requirements. *The National Law Journal,* September 30, pp. 26–27.

Stepke, K. (1987, September/October). How to develop effective (and legal) personnel tests. *Legal Administrator, 6,* 28–34.

Subkoviak, M.J. (1988). A practitioner's guide to computation and interpretation of reliability indices for mastery tests. *Journal of Educational Measurement, 25(1),* 47–55.

Swezey, R.W. (1981). *Individual performance assessment: An approach to criterion-referenced test development.* Reston, VA: Reston Publishing Company.

Thompson, D.E., & Thompson, T.A. (1982). Court standards for job analysis in test validation. *Personnel Psychology, 35,* 865–874.

U.S. Equal Employment Opportunity Commission, U.S. Civil Service Commission, U.S. Department of Labor, & U.S. Department of Justice. (1978). Uniform guidelines on employee selection procedures. *Federal Register, 43,* 38290–38309.

U.S. Equal Employment Opportunity Commission, U.S. Civil Service Commission, U.S. Department of Labor, & U.S. Department of Justice. (1979). Adoption of questions and answers to clarify and provide a common interpretation of the Uniform Guidelines on Employee Selection Procedures. *Federal Register, 44,* 11996–12009.

U.S. Equal Employment Opportunity Commission, U.S. Civil Service Commission, U.S. Department of Labor, & U.S. Department of Justice. (1980). Adoption of additional questions and answers to clarify and provide a common interpretation of the Uniform Guidelines on Employee Selection Procedures. *Federal Register, 45,* 29530–29531.

Index

Administering the test
environmental factors, 89–90
giving and monitoring, 90–92
preparing for, 81–82
Adverse impact, 201–204
hiring ratios, 201–202
illustrated, 202
record keeping and, 209
Affirmative action, 205–206
Agreement coefficient
calculating, 140, 177–178
description of, 177
equation for, 140, 177
inter-rater reliability and, 137
level of, 178–179
limitations of, 184
test-retest reliability and, 177
Alexander, R. A., 206
Analysis level, Bloom's Taxonomy, 37
Angoff-Nedelsky method. *See* Conjectural approach
Application level, Bloom's Taxonomy, 36–37
Assessment, 8
Attitudes, 40

Average correlation coefficient error, 151
See also Fisher's Z coefficient
Average score, 107

Ballew, P. J., 200
Barrett, G. V., 206
Barrett, R. S., 45–46
Behaviorally anchored numerical scale, 77
Bell curve. *See* Normal curve
Berk, R. A., 199
Bloom, B. S., 52
Bloom's levels, 52, 54–57, 60–61
cognitive classifications, 52
See also Bloom's Taxonomy; Test items
Bloom's Taxonomy, 35–38, 51–52
cognitive levels of, 35–37, 52, 54–57
validating a hierarchy and, 37–38
Borderline decisions, 128–129
reducing classification errors and, 131–132
standard error and, 129–130

215